W9-BNS-626

SAY AGAIN, PLEASE
GUIDE TO RADIO COMMUNICATIONS

BOB GARDNER

FOURTH EDITION

Aviation Supplies & Academics, Inc.
Newcastle, Washington

Say Again, Please – Guide to Radio Communications
Fourth Edition
by Bob Gardner

Aviation Supplies & Academics, Inc.
7005 132nd Place SE
Newcastle, Washington 98059-3153
www.asa2fly.com

The flight and radio talk examples used throughout this book are for illustration purposes only, and are not meant to reflect all of the possible incidences and communications that may occur in actual flight, nor does the author suggest by using existing facilities that the flight example given covers all possible parameters of an actual flight to or from those facilities. The airport photographs and chart excerpts are not for navigational purposes; refer to the current charts and Airport Facilities Directory when planning your flight.

Printed in the United States of America

2012   2011   2010   2009   9   8   7   6   5   4   3   2   1

ASA-SAP-4
ISBN 1-56027-760-2
        978-1-56027-760-6

Photo and Illustration Credits: Aerial views of Washington State airports, courtesy Washington State Department of Transportation, Aviation Division; p.viii, Jim Fagiolo; p.2-2, p.2-3, Bendix/King; p.2-4, courtesy Garmin; p.2-5 through 2-12, Telex Communications, Inc.; p.2-10 (left), ASA, Inc.; p.2-12 (left), Sigtronics; p.2-13 (top) King Silver Crown; p.2-13 (bottom), Terra; p.2-15, Narco Avionics; p.3-2, 3-4, 3-6, 6-1, 10-3, Bob Gardner; p.3-11, Henry Geijsbeek; p.6-9 Olympia airport guide, courtesy Airguide Publications, Inc. **Cover Photos:** air traffic control tower, George Clerk/iStockphoto.com; pilot, ColorBlind Images/Blend Images (RF)/Jupiterimages.

Library of Congress Cataloging-in-Publication Data:

Gardner, Bob
        Say again, please : guide to radio communications / by Bob Gardner.
                p.   cm.
        "A Focus series book."
        Includes index.
        1. Radio in aeronautics. I. Title.
        TL693.G34 1995
        629.132'51 – dc20   95-22588
                                                                CIP

Contents

3 A Matter of Procedure

4 Class G Airspace

5 Class E Airspace

6 Class D Airspace

7 Class C Airspace

8 Class B Airspace

9 Class A Airspace

10 Automated Flight Service Stations

11 The IFR Communicator

12 Now That You Know the System...

Appendix A Communications Facilities

Appendix B Airspace Definitions

Appendix C Clearance Shorthand

Glossary

Index

Sectional Chart Foldout

About the Author

Bob Gardner has long been an admired member of the aviation community. He began his flying career as a hobby in Alaska in 1960 while in the U.S. Coast Guard.

Bob's shore-duty assignments in the USCG were all electronic/communications based. He served in the Communications Division at Coast Guard Headquarters and was Chief of Communications for the Thirteenth Coast Guard District. He holds a Commercial Radiotelephone Operator's license and an Advanced Class Amateur Radio Operator's License.

By 1966, Bob accomplished his Private land and sea, Commercial, Instrument, Instructor, CFII and MEL. Over the next 16 years he was an instructor, charter pilot, designated examiner, freight dog and Director of ASA Ground Schools.

Currently, Bob holds an Airline Transport Pilot Certificate with single- and multi-engine land ratings; a CFI certificate with instrument and multi-engine ratings; and a Ground Instructor's Certificate with advanced and instrument ratings. In addition, Bob is a Gold Seal Flight Instructor, has been instructing since 1968, and was awarded Flight Instructor of the Year in Washington State. To top off this impressive list of accomplishments, Bob is also a well-known author, journalist and airshow lecturer.

He can be contacted on the Internet at bobmrg@comcast.net.

Books by Bob Gardner:
> *The Complete Private Pilot*
> *The Complete Private Pilot Syllabus*
> *The Complete Multi-Engine Pilot*
> *The Complete Advanced Pilot*

Software and Audio Review by Bob Gardner:
> *Communications Trainer*

Introduction

We live in a technological age. It is possible to fly without radios or electronic aids to navigation and rely solely on the Mark I eyeball, but there is no question that safety is enhanced when pilots can locate one another beyond visual range. The avionics industry continues to provide pilots with improved products which make communication easier and more reliable, but technology alone is not enough—the user must feel comfortable with the equipment and the system.

We all feel comfortable with the telephone, and an increasing number of pilots feel comfortable with radios that operate in the citizen's or amateur radio bands. However, if there is a controller on the other end of the conversation many pilots freeze up. The goal of this book is to increase your comfort level when using an aircraft radio by explaining how the system works and giving examples of typical transmissions.

A brief word of explanation. I am a flight instructor, and flight instructors talk, and talk, and talk. It is impossible for me to shut off my flight instructor instincts and convert myself totally into a writer. You will pick up on this right away because I repeat myself. Over 30 years of instructing I have learned that if something is repeated in different contexts it will be remembered, so you can count on the same information showing up in more than one chapter. That is not sloppy editing or carelessness, it is good instructional technique. Also, some types of airspace change classification when the tower closes down or the weather observer goes home—there will be some overlap as I discuss each situation in the chapter on each type of airspace.

Conventions

I will not spell out numbers in this text; the AIM says that numerals are to be pronounced individually: 300 is spoken as "three zero zero," runway 13 as "runway one three," etc. I know that I can count on you to make the mental conversion. Altitudes are handled differently, as you will learn in Chapter 3. Also, controllers do not say "degrees" when assigning courses and headings, so neither will I.

In radio communication, the different classes of airspace are spoken as their phonetic equivalents (again, *see* Chapter 3), without the word "class":

> **"Cessna 1357X is cleared to enter the Charlie surface area..."**

In the text, however, they will be referred to as Class B, Class G, etc.

Editor's Note

The examples of radio talk between pilots, controllers and other communications facilities in this text are printed in a bold and italics, non-serif typeface. These are also identified by small labels, which are sometimes abbreviated, as visual aids to the reader to show who is talking. Definitions for these labels can be found in Appendix A, "Communications Facilities."

Example:

PILOT *"Cessna 1357X requests runway 23."*

Acknowledgements

The author wishes to acknowledge the assistance of the following experts in reviewing the text for accuracy and completeness:

Suzanne Alexander, Manager, Boeing Field Tower
Jim Davis, Plans and Procedures, Seattle-Tacoma TRACON
Terry Hall, American Avionics, Seattle
Mike Ogami, Seattle Automated Flight Service Station

Note about the examples used in this book:

The National Aeronautics and Space Administration (NASA) commissions contractors to search the NASA database for lessons to be learned from accidents and pilot reports. Also, NASA publishes *Callback*, a free monthly newsletter that provides its subscribers with selected incidents from the Aviation Safety Reporting System (ASRS). Except for those few cases where I received an anecdote directly from an ATC controller, the examples in this book come from NASA sources.

If you want to receive Callback, simply send your address to ASRS, Box 189, Moffett Field, California, 94035 or view online at:

http://asrs.arc.nasa.gov/callback_nf.htm

If you want to "hear" and see this book in action, check out the *Communications Trainer* (order number ASA-ESAP) software product, which also includes an Audio Review so you can listen to many more examples of communication exchanges on your home or car stereo.

THE ABCS OF COMMUNICATING

The Pilot-Controller Partnership For Safety

Aviation communication is a team effort, not a competition between pilots and controllers. Air traffic controllers are just as anxious as you are for your flight to be completed safely. They will cooperate with you whenever they can do so and still remain *consistent with safety*. They are not the equivalent of the stereotypical law enforcement officer just waiting for you to do something wrong. They hate paperwork as much as anyone, and filing a violation against a pilot starts an avalanche of forms and reports. On the other hand, they have a tremendous amount of responsibility and can be severely overloaded with traffic; that means you can't expect a controller to ignore everyone else in order to give you special treatment.

Doing Things by the Book

The controller's actions are bound by FAA Handbook 7110.65, the *Air Traffic Control Handbook*. This publication tells controllers exactly what phraseology to use in virtually every situation, and woe to the controller who has had a slip of the tongue when he or she sits down with a supervisor to jointly monitor tapes during a quarterly evaluation. That is not to say that the controller operates in a procedural straitjacket. If you don't understand what a controller has said, or do understand but don't know what you are being told to do, just say "I don't understand," or words to that effect. The controller won't be out pounding the pavement, since the intent of the communication was to extend a helping hand and make your life a little easier.

As a pilot, you do not have a manual of canned phrases that are expected to meet every situation. The *Aeronautical Information Manual* contains a section on communication procedure, and if you read it (and you should) you will receive guidance on the best way to get your message across to the controller.

Both the *Aeronautical Information Manual* (AIM) and the *Air Traffic Control Handbook* contain the Pilot/ Controller Glossary. The intent of the Glossary is to ensure that certain words have the same meaning to both the pilot and the controller. Before you ask your instructor a question like "What does 'resume own navigation' mean?" look it up in the Pilot/ Controller Glossary. There are very few terms used in normal aviation communication that do not appear in the Glossary.

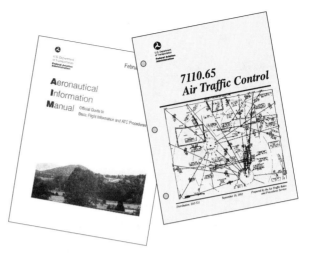

Figure 1-1. AIM and ATC Handbook

An historical sidelight: The Pilot/Controller Glossary didn't exist before 1974. It became apparent only after a major airline accident that some phrases meant one thing to controllers and something entirely different to pilots, and the glossary was born. A very good reason for you to familiarize yourself with the P/C Glossary in the AIM.

Can't We All Just Get Along?

An important part of the teamwork concept is negotiation. Many pilots, both novices and old hands, think that a directive from an air traffic controller must be obeyed without question. Those pilots have forgotten that the Federal Aviation Regulations make the pilot-in-command of the airplane solely responsible for the safety of the flight. A controller cannot direct you to do something that is unsafe or illegal. You must remember that you are almost always in a better position to determine the safety of a given action than is the controller.

For example, let's assume that you are flying in Class B airspace (to be defined later). In that type of airspace the controller can give you specific altitudes and/ or headings to fly; you are required by 14 CFR §91.123 to comply with those clearances. When the controller says "Turn right to 330" and you can see that to do so would take you too close to a cloud, it becomes your responsibility to say "Unable due to weather." After all, the controller can't see clouds on the radar screen and has no way of knowing that you would be turning toward a cloud.

14 CFR §91.3 says that you are the final authority as to operation of your aircraft and this rule supersedes all others.

Another example: You have just touched down on the runway and the controller says "Turn right at the next taxiway." If you are rolling too fast to make the turn without wearing a big flat spot on your main landing gear and overheating the brakes, it is your responsibility to say "Unable." If you are really busy with the airplane, don't say anything until you can reach for the microphone without losing directional control.

Other situations where negotiation might be used include being assigned a landing runway that requires a lot of taxiing to get to your destination or, in light winds, a departure runway that takes you in a direction that you don't want to go. Simply say,

PILOT *"Cessna 1357X requests runway 23"*

(instead of runway 14, for example). If the change can be accomplished without affecting either your safety or that of other flights, your request will be granted. There are almost as many exceptions to the rules as there are rules, but too many pilots simply go by the rules without attempting negotiation.

Mike Fright

We are all afraid of saying the wrong thing, especially when dozens of other people are listening. Aviation magazines frequently print stories of humorous communication mistakes or misunderstandings. In aviation, it is far more important to say something than to keep quiet and proceed into a potentially tight situation — especially when traveling at two miles a minute.

Call-in talk shows are quite common on both radio and television, and the callers are in the same situation as you are when you pick up the microphone in your airplane as a "first-time caller" — thousands of people will be able to hear their "er's" and "uh's." The difference is that their safety and that of others does not depend on their making that call — yours does.

Technobabble Not Spoken Here

Use plain English. "Tell me what you want me to do" might not appear in the AIM, but if it is necessary to use that phrase, it gets the job done. The following suggestion will be repeated later more than once, because it is important: Listen to your radio. Other airplanes will be talking to ATC, getting weather reports, or communicating with advisory services. The information they are receiving might be useful to you and make it unnecessary for you to make a transmission (or allow you to drastically shorten your transmission). Go to any small airport (one without a control tower) with a VHF receiver that covers the aviation frequencies and just monitor the airport's Common Traffic Advisory Frequency (CTAF) — ask one of the local pilots if you aren't sure what the CTAF for that airport is. You will hear a dozen airplanes reporting that they are landing or taking off on runway 14 (for example), and then a strange voice will come on the frequency and ask "What runway is in use?" That pilot hasn't learned to listen.

That VHF receiver is your best source of information on how to communicate as a pilot. Get a copy of the Airport/Facility Directory (A/FD) for your area and look up the frequencies that are used by the local airports and air traffic control facilities. Look in the back of the directory for Air Route Traffic Control Center (ARTCC) frequencies, then tune in and listen to how the airliners communicate when en route. You will hear lots of good examples and a few alarmingly bad examples. You may not be able to hear both ends of the communication unless you live within line-of-sight distance of the ground station's antenna, but a visit to a local tower-controlled airport will eliminate that problem.

When you are surfing the web, spend some time at www.liveatc.net. On your computer, you will be able to listen to controller-aircraft traffic at a number of facilities nationwide and internationally.

While you are at your computer, go to www.faa.gov and click on "Regulations and Guidance" in the right column. Then click on "Orders and Notices." That will lead you to FAA Order 7110.65, the *Air Traffic Control Handbook*. This directive tells controllers what to say and how to say it, and they are required to follow its dictates. This is important to you because you will see that controller transmissions follow a fixed format for each situation; only things like headings, altitudes, and facility names change. With this in mind, you will know what to expect in each situation. However, if it becomes apparent to the controller that the approved phraseology is not getting through to you, he or she is free to use plain language. By the same token, you are free to say, "I don't understand what you want me to do" if that is the case. Most of the ATCH will not apply to you, but read it anyway…it is a treasure trove of information.

You might want to take a look at www.asf.org/askatc. This site offers pilots the opportunity to ask controllers any and all questions about communications. You do not have to be an Air Safety Foundation member to access this site. The ASF also has a free program called "Say it Right," available at www.asf.org/courses. In it are illustrated many, if not all of the lessons in this book.

Licensing

Federal law does not require U.S.-registered airplanes to have a Federal Communications Commission (FCC) radio station license unless international flight is contemplated—an FCC license is not required for U.S. operations.

To use an installed aircraft radio in the United States, you don't have to have any kind of operator's license. Travel to Canada or Mexico, however, and you will be expected to carry a Third Class Restricted Radiotelephone operator's license. Any FCC office can issue one if you promise to avoid saying bad words on the air. An FCC commercial license is even better than a Third Class Restricted license, but an amateur radio license is no good on aircraft frequencies. Keep in mind that you won't even get into a hassle about an operator's license unless you are outside of the good old U.S.A. and subject to another country's laws. To my knowledge, neither the Canadian Mounties nor the Mexican Federales are enforcing this requirement.

Handheld transmitter-receivers (to be called transceivers from here on) are very popular for emergency use. If you are flying an airplane with its own station license, that license covers the handheld. You will use the airplane registration number as an identifier. If you want to use a handheld in an airplane without a station license, however, you will have to apply to the FCC for a "mobile" license, which will assign an identifier to the handheld—something like "N12345MOB."

Don't look to the Federal Aviation Regulations for anything about how radios are to be used—that is the sole province of the FCC.

Hello, Operator?

Cellular phones are becoming more and more popular, and many pilots use their cellular phones in business. What could be better than calling that hot prospect from 12,500 feet and setting up a lunch date? The hard fact is that doing so would violate FCC (not FAA) regulations. The cellular phone system is designed to be used on the surface, where callers are automatically switched from one small cell to another as they travel. From 12,500 feet or any other altitude, the signal from your phone might activate a dozen cells at once, and that is frowned upon by the FCC. To make matters worse, when you use your cell phone it transmits enough identifying information (for billing purposes) for the FCC enforcement officers to easily prove a violation.

There are approved airborne telephones—ask at your local avionics shop. One entrepreneur (AirCell) has received FCC approval of a cellular network specifically for airborne cell phones. Their equipment and service is more expensive than your regular wireless service.

Can you use your cell phone on the ground to call for gas or file a flight plan? Absolutely.

By the way, leave your cell phone in your luggage while on an air carrier. Even if you want to use it on the ground, you may find that the flight attendants will not permit you to do so. The regulations make the captain the final authority on the use of portable electronic equipment, and there are documented instances of cell phones interfering with air carrier avionics while in the receive mode.

Approved by OMB
3060-0049
Expires 5/31/97
See reverse for
public burden.

FEDERAL COMMUNICATIONS COMMISSION

FOR
FCC
USE
ONLY

FCC 753 - Application for Restricted Radiotelephone Operator Permit

Read instructions on reverse before completing this form. No examination is required.

1. Name :   Last   First   Middle Initial

2. Address to which permit should be mailed - Number and Street   City   State   ZIP Code

3. Payment Type Code   P A R R   4. Fee Due $ FOR FCC USE ONLY   5. Date of Birth Use numeric
MONTH   DAY   YEAR

I CERTIFY THAT:
- I am legally eligible for employment in the United States.
- I can keep at least a rough written log.
- I am familiar with the provisions of applicable treaties, laws, and rules and regulations governing the radio station which I will operate.
- I can speak and hear.
- I need this permit because I intend to engage in international flights or voyages (Aviation and Marine Services only).
- The statements made on this application and any attachments are true to the best of my knowledge.

By checking "YES", the applicant certifies that he or she is not subject to a denial of federal benefits, that includes FCC benefits, pursuant to Section 5301 of the Anti Drug Abuse Act of 1988, 21 U.S. Code 862.

DOES APPLICANT SO CERTIFY? ☐ YES ☐ NO (You must check one.)

Willful false statements made on this form are punishable by fine and/or imprisonment (U.S. Code, Title 18, Section 1001), and/or revocation of any station license or construction permit (U.S. Code, Title 47, Section 312(a)(1)), and/or forfeiture (U.S. Code, Title 47, Section 503).

SIGN HERE ☞   Date

FCC Form 753 PART 1 August 1994

FCC Form 753 - PART 2
August 1994   DO NOT MAIL THIS PART OF THE FORM - IT IS YOUR TEMPORARY PERMIT   Approved by OMB
3060-0049
Expires 5/31/97

FEDERAL COMMUNICATIONS COMMISSION
TEMPORARY RESTRICTED RADIOTELEPHONE OPERATOR PERMIT

If you need a temporary Restricted Radiotelephone Operator Permit while your application is being processed, do the following:
- Complete Part 1 of this form and mail to the FCC.
- Complete this part of the form and keep it.

This permit is valid for 60 days from the date Part 1 of this form was mailed to the FCC.

You must obey all applicable laws, treaties, and regulations.

Read, Fill in Blanks, and Sign:

Name

Date FCC Form 753, Part 1 was mailed to FCC

I CERTIFY THAT:
- The information in Part 1 is true.
- I have signed FCC Form 753, Part 1 and mailed it to the FCC.
- I have never had a license suspended or revoked by the FCC.

Signature   Date   If you cannot certify to all of the above, you are not eligible for a temporary permit.

FCC Form
August 1994

(partial underlying page, left)

WHO SHOULD USE THIS FORM
You should use this form if you
are considered, for the purpose
you are NOT eligible for employ

→ you hold a
need to

→ you hold
for oper

then use FCC Form 755 to

FEE AND MAILING INS
FCC 1070-R for curre
Branch at (717) 337-

NOTICE TO
The solicitation of personal in
use the information provided in
law enforcement purposes. If
addition, all information prov
not provided, processing of t
foregoing notice is required b

Public reporting burden f
instructions, searching exi
information. Send comme
the burden to Federal Co
Management and Budget

TERMS AND CO
It is your resp
you operate
station lice
Communic

As a licens
- w lif
- tran
- fals

- transmit without authorization, divulge contents, substance ... transmissions inter...
here to, the existence, content, other than transmissions relating to ships, aircraft, vehicles, or persons in distress; or
communications by radio, other than
transmissions relating to ships, aircraft, vehicles, or persons in distress; or
or citizens band radio operator.

Figure 1-2. FCC license application, required only for operations outside the U.S.

Look At All Those Knobs and Buttons!

When you had your first flight lesson, your instructor probably went to great lengths to tell you how to read the instruments and how to use the controls. During that and subsequent lessons the proper use of the radios and their controls were most likely glossed over. After all, what can be difficult about selecting a frequency and picking up a microphone? If the airplane had an audio control panel or an intercommunication system, it is possible that the instructor set them up and their use was taken for granted (my apologies to all instructors to whom these generalities do not apply).

If you did all of your training in one airplane, let's say a Cessna 152, the first time you got into a different 152 the odds are that the radio setup was different. Or when you got your license and checked out in a larger airplane, its radio installation was different than the one in the 152. At a fixed-base operation (FBO) that uses airplanes leased back from individual owners, the radio installation in each airplane will be tailored to meet the requirements of the owner, not the FBO.

Communications hardware can be confusing. It would be impossible to describe all of the possible combinations you might encounter, and if I attempted to do so my discussion would be out of date in six months. Another complicating factor is conversion of old, mechanically tuned radios to the new digital displays. It's hard to stay ahead of avionics technology. I can, however, give you a general overview of communications hardware.

Transceivers

A typical aircraft nav-comm is shown in Figure 2-1; the navigation function will be ignored in this text. Your aircraft radio transmits and receives on frequencies between 118.000-136.975 MHz (the display of 136.97 in Figure 2-1 is 136.970; you get 136.975 by pulling the knob marked "PULL 25K). There is no .005 display until you get into really sophisticated airplanes with cathode-ray tube displays. Whether or not it has .025 MHz spacing will depend on its age.

Digital radio is coming...not digital display, but voice transmission converted to digital form. Channel spacing will be 8.33 kHz. It will be used initially in the high altitude structure, where only instrument pilots fly, but will migrate down to lower altitudes over the years. Don't panic: your radio will be able to handle both old and new modes.

Other than selecting the correct frequency, there is little you can do to a transmitter. Radio manufacturers have made great technological advances since microprocessors became available, and one improvement with a large impact in this discussion is frequency storage. Note that two frequencies are displayed on the comm side of Figure 2-1: the top one is the active frequency, and the bottom one is the stand-by frequency, and if you push the button on the left, 136.97 would jump up to the top and become the active frequency.

If the radio in the airplane you are flying has mechanical frequency display — white numbers on a black background that flip over like a speedometer as you turn the frequency selection control — you don't have storage capability. With this mechanical type of frequency selection you can turn the frequency selector clockwise from 118.000 to 136.900, but a mechanical stop prevents you from going

Figure 2-1. Nav-comm transceiver

any further. With digital tuners you can go right through 136.900, and the display starts over again at 118.000. This is not only a timesaver but saves wear and tear on the mechanical linkages as well.

Digital displays, those orange-on-black, green-on-black, or black-on-white segmented numerals that have become so common in everyday life, tell you that your transmitter/receiver probably has frequency storage and flip-flop capability. The simplest arrangement is typified by the KY-96A, which has no navigation receiver to confuse the issue (Figure 2-2).

A transceiver with flip-flop capabilities such as the KY-96A allows you to select a new frequency on the standby side while transmitting and receiving on the frequency shown on the active side. You might have the tower frequency selected (or stored) on the standby side while using the ground control frequency on the active side. Radios similar to the KY-96A can store many frequencies (nine, in the case of the KY-96A) to be called up at the push of the CHAN button. You can always stay one step ahead if you have some idea of what the next frequency to be selected will be, and you can cycle through the stored frequencies with that button.

Another variation on frequency storage doesn't use a flip-flop but lets you choose three or four frequencies that are selectable by a switch; at an airport in Class D airspace, you might have the ATIS (Automated Terminal Information Service — to be defined later) in position 1, ground control in position 2, and the tower frequency in position 3. As soon as you have received the ATIS information on position 1 and have switched to position 2 for taxi instructions, you can retune position 1 to departure control, UNICOM, or whatever frequency you plan to use after leaving the Class D airspace.

Figure 2-2. KY-96A transceiver

With two transceivers, use #2 for ground control or clearance delivery, just to ensure that it works. Then use #1 for all further transmissions and keep #2 for listening to weather, ATIS, the emergency frequency, etc. Alternating between #1 and #2 for transmitting just increases the chances of error.

GPS/Communication Devices

As the Global Positioning System (GPS) and moving maps have become more pervasive in general aviation, manufacturers have become very innovative. Combination GPS navigators/VHF transceivers are now available for virtually every kind of airplane (Figure 2-3). The big plus from the pilot's seat is that when you program your destination waypoint into the GPS (it already knows where you are, of course), its NAV FREQ page loads the appropriate frequencies for the direct route in order of expected use; approaching the destination it does the same thing. Select a frequency with a cursor and it is automatically set up in the communications transceiver. For radar flight following (Chapter 5), the nearest Air Route Traffic. Control Center frequency is on another page of the electronic display.

Figure 2-3. The GNS 430 is a GPS/VHF nav-comm unit.

Squelch

The squelch control permits you to fly along in blessed silence without having the continuous roar of static in your ears. The only time any sound will reach your ears is when a receiver signal is strong enough to overcome the squelch setting or "break squelch."

Many newer radios have an automatic squelch which is set by your radio shop. Older radios will have a squelch knob, and this is how you use it: Turn the squelch knob until static assaults your eardrums, set the volume to a comfortable level, then turn the squelch knob until the static is silenced — the correct setting is right on the edge of squelch break. If you turn the squelch knob beyond that point, you establish a squelch level that is too high for incoming signals to exceed before you can hear them.

Radios with automatic squelch will have a selector switch with a "pull-to-test" position that bypasses the squelch control (Figures 2-1 and 2-2). This is valuable for setting a comfortable audio level, but it also has a more practical use. What if you are 75 miles away from your destination airport and want to check on the weather and the runway in use? Chances are that the ATIS signal will be too weak to break squelch at that distance. However, if you select the "test" position you may be able to decipher the ATIS through the background of static.

Many newer radios have an illuminated "T" next to the transmitter frequency when you are talking on that radio. It tells you that your transmitter is working, of course, but it is also a warning that your transmitter may have been keyed accidentally.

The microprocessor brought with it miniaturization, and miniaturization means that it is possible to put more avionics goodies in a panel than were possible in the past. This is good. Solid-state devices don't put out as much heat as vacuum tubes did, and that is also good. However, solid-state devices such as transistors and microprocessors are very temperature-sensitive, and that is bad. It means that your radio stack will probably require one or more cooling fans or ram air from outside. There is also the possibility that your solid-state equipment contains a thermal shutoff device that will remove the power from some of your avionics if the temperature gets too high. Talk this over with someone from the avionics shop that services the airplane so that you know which pieces of equipment might be affected. It's much less expensive to install a cooling fan than it is to have your radios in the shop constantly.

Mikes

Unless you are an amateur radio operator, a user of the Citizen's Band, or a radio announcer, it is likely that the microphone in your trainer airplane will be the first mike you have used. All microphones perform the same function — turning your voice into electrical waves — but they differ in how they accomplish this task.

The microphone found in most trainers is a carbon mike (Figure 2-4). It works exactly like a telephone handset

Figure 2-4. Carbon microphone

in that it contains a little canister of carbon granules. One side of the canister is a thin diaphragm that vibrates as the sound waves from your mouth hit it. This vibration changes the electrical resistance of the granules and thereby varies the current flowing through it. That's as technical as we need to get. That semi-technical explanation gives you some insight into how you can get the most out of your relatively inexpensive carbon mike.

First, you have to be sure that you hold the mike directly in front of your mouth so that the amplitude of your voice's sound waves is fully utilized—you can't talk across a carbon mike or speak into one from several inches away and expect great results. Most carbon mikes have a little protrusion that you should press against your lip to make sure that the granules get all shook up as you speak. Talking into the back of the microphone accomplishes absolutely nothing, yet every year, lots of pilots try it.

Second, it is possible for the granules to stick together (humidity? cigarette smoke?) and not vibrate as much as they should. This is not a common problem, but if you are holding the mike properly and speaking into it distinctly and still get reports that your voice is weak, gently rapping the mike on a hard surface might help. Please don't do this while keying the microphone—you don't want to damage other pilots' eardrums. Carbon mikes can withstand a certain amount of physical abuse; that explains why they are so often found in trainers. The other side of the coin is that the granules can vibrate only so much. When you yell into the microphone your voice doesn't get louder at the receiver end, it just gets distorted. If you live in New York and call California on the telephone, do you yell because your voice has to travel so far? Yelling into any kind of microphone is counterproductive.

Stepping up in class, there are dynamic mikes (Figure 2-5), crystal mikes, electret mikes, and condenser mikes. There may be a few more by the time you read this book. These share a common characteristic: they don't react well to physical abuse. It is that same sensitivity that makes them valuable in a noisy cockpit. It is not important for you to know how these mikes work. Suffice it to say that speaking into one in a conversational tone will produce a readable signal at the receiver.

Figure 2-5. Dynamic microphone

All handheld microphones have a push-to-talk (PTT) switch (in Figure 2-5, the PTT is on the front and you talk into the top). True to its name, you have to push the switch ("key the mike") to transmit, and release it to receive. A common error among neophytes who expect a radio to act like a telephone is to push the button and then hold it down after completing their transmission. You can't hear while you are transmitting because the receiver is blocked, so maybe they should be renamed PTTRTR switches: Push to talk, release to receive.

As you sit in the cockpit of the typical single-engine airplane you are surrounded by noise. The engine does its noisy thing, the propeller sends waves of air back to vibrate the windshield, and if there are any airleaks around the doors or windows they will add a noise like a banshee's wail. On the receiver end, the controller or pilot must somehow distinguish your voice from the background noise. Fortunately, almost all microphones are designed to be noise-canceling. That is, they are extremely sensitive in the direction of your mouth and they are designed to internally cancel noise coming from anywhere else. That's another reason for you to be sure that the business end of the microphone is very close to your lips.

Speakers

Almost every airplane has a speaker installed in the headliner so that people in the front seats can hear. More often than not, it is a very small speaker and not of the highest quality. If that speaker fails you will have a very real loss of two-way communication capability. One of the things you should check before takeoff is the presence of headphones that you can use in case the speaker dies. They don't have to be the most expensive headphones on the market, just something to get you back on the ground safely. Don't bring your stereo headphones from home—they won't work.

Silence from the speaker could be caused by failure of the audio amplifier required to drive it. This amp could be a separate unit, included with the audio switch panel, it could be part of your nav-comm receiver, or it could be a stand-alone unit. The important thing to remember is that the speaker can suddenly fall silent, and you should have a backup method of listening to your communications radios. When you use the headphone jack and/or throw the switches on the audio panel from SPKR to PHONE, you bypass the audio amplifier, thus the need for headphones to back up the speaker.

Headphones

Figures 2-6 and 2-7 show two types of headphones, and you should know that there are several other types not illustrated. Do you like to hear slightly muffled engine noise or do you want blessed silence? The difference might be the several hundred dollars you pay for noise-canceling headphones. Do you want to hear ATC with one ear while monitoring the passengers, and the cabin speaker with the other? The EarSet® (Figure 2-6) is one solution; you can get a separate boom mike that hooks onto your sunglasses. Passengers may be concerned when the pilot puts on a big headset (Figure 2-8) and effectively goes into seclusion during the flight: "Did you hear that noise, Mabel? The

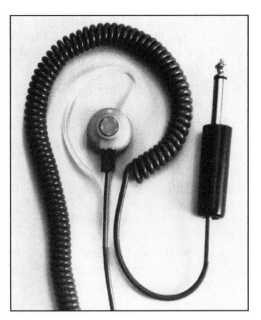

Figure 2-6. EarSet® receiver with earloop

pilot couldn't have heard it and I think the wing is falling off!" It's also a major drag for the instructor when the student is using muff-type headphones and there is no mike or phone connection on the right side of the cockpit. They will be able to communicate through gestures and shouting, but it won't be easy.

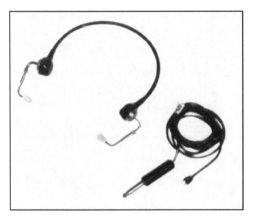

Figure 2-7. Lightweight headphones

The amount of cushioning around the earpiece, and therefore the amount of cabin noise/conversation you can monitor, varies quite a bit. Try several types, but be warned — over a flying career you will probably buy two or three headsets of different types. Most headsets have their own volume controls, so if each occupant of a pilot seat is wearing a headset the volumes can be controlled individually. With a volume control on the receiver, one on the intercom, and another on the headset, there should be sufficient control to satisfy everyone.

Boom Microphones

You won't be flying very long before you decide that reaching for a handheld microphone and returning it to its holder is a lot of trouble and distraction. You don't solve this problem by sticking the mike under your thigh where you can get at it easily. More often than not, pilots who do this key the mike by shifting around in their seat, and for the next several minutes every word said in the cockpit is transmitted to hundreds of ears. While they are sitting on the mike and blithely transmitting their life story to the world at large, other pilots are unable to contact controllers, controllers are unable to contact other pilots, and a large number of people are getting very angry at the thoughtless pilot. Sometime during the conversation someone is bound to say the aircraft's N-number, making it all too easy for the FAA, the FCC, and an army of angry pilots to identify the miscreant.

The best way to solve the distraction problem is with a boom mike (Figures 2-8 through 2-11). If you are just beginning your flight training and your instructor insists that you use a boom mike, you have a very smart instructor. Telephone operators and onstage performers often use boom mikes. In every case, the business end of the mike is positioned either directly in front of the mouth or very

Figure 2-8. Headset with volume control and boom mike

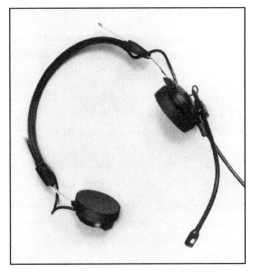

Figure 2-9

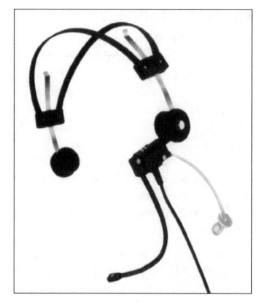

Figure 2-11

Figure 2-10

close to that centered position, and surrounded by a foam plastic windsleeve to keep extraneous noises from being picked up. Your airplane's boom mike may have a similar sleeve—almost all aviation boom mikes are extremely directional.

The boom portion of a boom mike can be swiveled and adjusted to put the mike right where you want it. One word of caution: Before you bite into an apple or a sandwich, swing the mike out of the way or that first bite will be mostly plastic. Another word of caution: If someone calls you while you are eating lunch and the boom mike is somewhere up near your forehead, swing it back into position before you key the mike and reply. I speak from experience.

Earphones with foam covers that fit over the ear but do not enclose it (Figure 2-9), and earpieces that actually fit into the ear canal (Figure 2-11) should be considered. Some pilots like to have one ear covered while they listen to ATC, while monitoring the cabin speaker, the passengers, and engine sounds with the other ear (just about any kind of earplug will protect your hearing while you do this). Your best bet is a visit to the local avionics shop where you can try all kinds of headsets to determine which suits you best.

You will be able to hear your own voice very faintly when transmitting while using a headset. This is called "sidetone," and if you don't hear it you may not be transmitting.

If the boom mike is continually in front of your mouth, where is the PTT switch? One of two places. If the owner of the airplane has had one installed, the PTT switch will be on the horn of the yoke, either directly under your thumb or on the front side under your index finger. Very convenient. But what if you are a renter who shows up at the airplane with a headset in a bag only to learn that the airplane does not have an installed PTT switch? Back to the avionics shop. PTT switches that can be temporarily attached to the yoke with hook-and-loop tape (Figure 2-12) are not too expensive and it doesn't make sense to buy a headset and not pop for a PTT switch, just in case. The headphone plug from your headset goes into the airplane's headset jack, the mike plug from your headset goes into the PTT adapter, and the mike plug from the PTT switch goes to the airplane's mike jack. Attach the PTT to the yoke where it is convenient to a finger with your hand in its normal flying position, and you are in business.

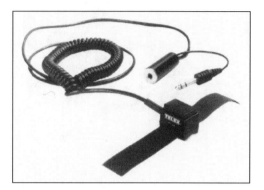

Figure 2-12. PTT switch

Intercoms

Intercommunication systems are becoming common in the general aviation fleet — the days of students and instructors trying to yell over the noise of a piston engine will soon be behind us. The ideal situation is an intercom that is permanently installed in the panel, but portable intercoms best fit the needs of renters. Once again, the intercoms in Figures 2-13 and 2-14 (on the next page) represent only a sampling of what is available.

In addition to the standard function of allowing both front seat occupants to either key the transmitter or converse with one another in normal tones, intercoms can bring the back seat passengers into the loop or provide them with musical entertainment. Another useful feature is a tape recorder output that allows you to tape your lessons and review them at your leisure. Some upscale intercom systems allow either the passengers or everyone on board to listen to music, automatically muting the music when a transmission is received.

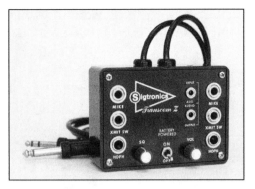

Figures 2-13 and 2-14. Portable intercoms

When you plug the mike and headphone connectors from your headset to the intercom's input jacks and then plug the intercom's connectors to the airplane jacks, the possibility increases that a loose connection might cause problems. If the receivers seem unusually quiet, or if you transmit and no one answers, check the connections first.

Some intercoms have a VOX (voice-operated) control, while with other systems it is automatic. VOX allows you to set a voltage level that permits your voice to be heard on the system but does not allow the system to be triggered by engine noise; it is used when the pilots want to talk to one another or to the passengers. It is adjusted by talking into the boom mike while adjusting the VOX until you can hear your own sidetone voice in your headset. Everyone plugged into the intercom system should be able to converse with this setting. With most systems, when either front seat occupant pushes the PTT switch, all of the mikes are live and a rear seat passenger might inadvertently ask ATC to pass the mustard. This possibility is eliminated when a "crew/passenger" switch is present and in the "crew" position. Your avionics shop will have a demonstration panel to show you how to use all of the unit's functions.

Audio Panels

Virtually every modern airplane has some form of audio switch panel (Figures 2-15 and 2-16) that allows you to switch between the cabin speaker and the headphones, or listen to two frequencies at once. (American Avionics will not install two radios unless an audio switch panel is included in the installation.) The switch panel will include a microphone selector switch for you to use when you have been talking

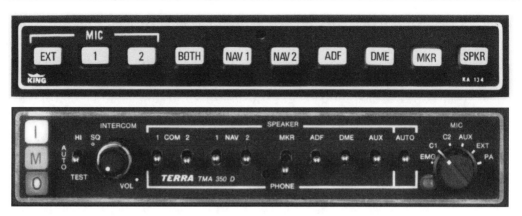

Figures 2-15 and 2-16. Audio panels

on your #1 radio and want to talk on #2. If you fail to switch transmitters you will, of course, transmit on the wrong frequency:

PILOT *"Dallas Tower, Baron 1014W ready for takeoff."*

GROUND *"14W, you're still on ground control."*

PILOT *"Sorry about that!"*

This is a mistake that even the most experienced pilots make, so don't feel bad if you do. But be more careful next time. Figure 2-15 provides for two transmitters, and when you select a transmitter its receiver is automatically selected. The SPKR push button is a toggle—it switches from speaker to headphone and back again. The BOTH position allows you to monitor the receiver of Comm 2 when you have the MIC 1 button pushed and are capable of talking and receiving on Comm 1.

The switches on panels like Figure 2-16 have a center "off" position, and if you are not careful to select either "speaker" or "phone," and leave the switch centered, you will not hear anything at all.

The panel in Figure 2-16 has an "auto" position. When the switches for both Comm 1 and Comm 2 are in the "off" position and the "auto" switch is in the speaker position, the output to the speaker is automatically switched from receiver 1 to receiver 2 when you change the transmitter switch from C1 to C2. If the "auto" switch is in the phone position, of course, the receiver output will go to the headphones. This makes it unnecessary for you to continually fiddle with the Comm 1 and Comm 2 switches.

There will be times when you will want to have (for example) Comm 1 set to the tower frequency while simultaneously listening to the ATIS on Comm 2. That can mean two signals coming into your ears at one time, but just keep your finger on the Comm 2 switch and you can turn it off when you are called by the tower.

The major advantage of audio switch panels is the ability to switch between navigation and communication radios without constantly fiddling with the volume and squelch controls. Once you have set the volume and squelch to a comfortable level you can leave those knobs alone and deal only with the switch panel.

Instructors differ on how to use the audio switch panel with the communications radios. One school of thought advocates using only the #1 radio for communications and using the #2 radio to monitor an ATIS, Flight Watch, 121.5 MHz, or other frequency you choose. This works quite well with radios that have a flip-flop capability because when you switch from one controller to another, the old frequency remains on display so that you can switch back if you are unable to contact the new controller.

The other school of thought advocates switching between the #1 and #2 radios every time you are handed off to another controller. This works well with mechanically-tuned radios that do not have flip-flop capability. When you are handed off from one controller to another and switch radios to accomplish the change, the "old" frequency is still mechanically displayed.

New pilots tend to spend a lot of time looking at the radios while changing frequencies, and this takes their eyes away from the important task of scanning for traffic. You should learn to tune your radios by counting "clicks." Once you have learned how many clicks of rotation it takes to change the frequency one megahertz, turn the frequency selector knob the number of clicks you estimate will get you to the one you want, while keeping your eyes outside of the cockpit. Then, take a quick glance at the radio panel to see how close you have come to the desired frequency. With practice, you will be able to change frequencies with hardly any need to look at the panel. Don't transmit until you have verified that you have selected the correct frequency.

Airplane cockpits are beginning to consist of all or mostly digital displays (*see* Figure 2-3). Many of them incorporate their own switching capabilities. Make sure that your instructor gives you a complete briefing on how everything works before you go off on your own.

Digital frequency displays do go blank occasionally, making it impossible for you to know what frequency is set in the window. Some manufacturers have an emergency mode that defaults to, say, 118.000. Then you can count the clicks to get to your desired frequency. If you have a digital display, this is something you want to be clear about before you fly with the equipment.

Transponders

Your airplane's transponder consists of a transmitter and receiver, so I can't ignore it in a book about communications. In normal VFR operations the transponder just sits there doing its job, blinking its little heart out while squawking code 1200. Every time it receives an interrogation from an air traffic control radar on the ground it sends out a coded reply. Its reply light may not blink too often when you are flying out in the boondocks where ATC radars are few and far between, but when you approach a busy terminal area it will blink continually.

At the controller's position your airplane will show up as an anonymous data block showing 1200 and, if you have Mode C, your altitude. When you call the controller for VFR flight following and are asked to push the IDENT button (don't hold it, just push it — it stays on for 20 seconds), your data block will intensify on the controller's scope.

If you are given a discrete transponder code to squawk and the controller has the scope set to display a full data block, your return will expand to include your callsign and groundspeed. Important note: When a controller asks you to squawk a discrete code, do not simply set the new code without a verbal acknowledgment; if you do not repeat the new code the controller might assume that you missed the transmission and be forced to take the time to transmit again. It would be a red-letter day at ATC if yours were the only aircraft the controller was working, and the controller cannot divert attention from other aircraft to see if you have responded correctly to the new assignment. Even worse, if the controller assigned code 4625 and you mistakenly set your transponder to 4325, your return would not show up on the scope at all and the controller would have to give you a call to find out what happened.

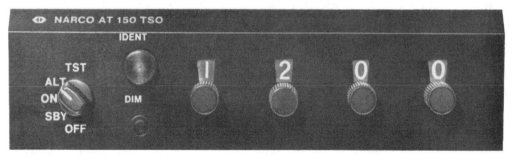

Figure 2-17. Transponder

The *Aeronautical Information Manual* says that you should not inadvertently squawk 7500, 7600, 7700, or 7777 while switching codes. This should be a fairly easy task; when switching from a 4 to a 1, for example, turn the selector counterclockwise to avoid passing 7 on the way. According to controllers, their radar sweep is so slow that they might not see the errant 7 anyway. Do not turn the transponder function switch to STBY while changing codes.

Code 7500 tells authorities that there is a hijacker on board; its intentional use by a VFR pilot on a pleasure flight is unlikely. Code 7600 reports to ATC a loss of two-way communications capability and is of primary interest to pilots on instrument flight plans. However, if you are receiving flight following services from ATC and your transmitter or receiver fails, the controller will worry about your failure to reply to advisories — it would be helpful if you would squawk 7600 until you land and can call a Flight Service Station on the telephone.

Of course, you might want to squawk code 7700 on purpose — it is the emergency code that alerts controllers to the fact that you have a problem of some kind. A few seconds spent in setting 7700 into your transponder (while changing tanks and looking for a flat spot on which to land) might save hours of searching by the authorities.

To conform with the regulations you should have your airplane's transponder ON at all times when you are flying in controlled airspace. This can be a real lifesaver, even if you do not avail yourself of VFR flight following, because ATC radar stores the tracks of all transponder returns. It is possible (not a sure thing, unfortunately) that if you turn up missing they will be able to replay the stored data to see where your 1200 squawk terminated. But if you fly so low that no radar can interrogate your transponder, there will be no display on the controller's scope.

"Mode A" on the transponder function switch provides only a target for the controller; "Mode C," or "ALT," adds your altitude to the data block on the ATC radar scope. Does the airplane you are flying have an altitude encoding altimeter? It should, but put the switch in the ALT position ("squawk" altitude) just in case. Like chicken soup for medicinal purposes, it can't hurt.

Where did this "squawk" stuff come from? During World War II, when the "Identification, Friend or Foe" system first used transponders, controllers and pilots called the system the "parrot" in an attempt to fool the enemy. Whether or not the enemy was fooled is subject to question, but today's request to "squawk standby" used to be "strangle your parrot." Only the squawk terminology has survived.

When you set your transponder to 7700, it tells air traffic controllers that you have an emergency of some kind by enhancing your data block on their scope. The FAA takes a very broad view of just what constitutes an emergency—a much broader view than that of most pilots. Any time you are apprehensive as to your safety you should ask for help. Too many pilots will wait until they are almost out of fuel before letting anyone know about their situation.

The best procedure is to ask for radar flight following. If you are in radio contact with ATC for traffic advisories, a word to the controller is all it takes to get help; your position is known and the controller is in direct contact with rescue agencies. No special communication procedure is required.

If you are not in voice contact with ATC, but are monitoring the ARTCC frequency for the area in which you are flying (listed in the Airport/Facility Directory), all you have to do is squawk 7700, pick up the mike, and say:

PILOT *"Center, Baron 1014W, emergency [or MAYDAY]"*

...followed by as much detail as you have time for. Believe me, they will be listening for your call, because of the way the 7700 squawk affects their radar scopes. The controller will then get the details of the situation and provide help as required. I strongly recommend that you either ask for flight following, or monitor a Center frequency whenever you fly cross-country.

The future of transponders is Mode S. Already available at high cost, Mode S transponders will provide an advanced collision avoidance capability. They also broadcast your aircraft's tail number, provide for datalink, and have many other advantages to offset their cost.

Figure 2-18. Mode S Transponder

Handhelds

Everything discussed so far has been either mounted in the instrument panel or plugged into a jack, and all of it requires power from the plane's electrical system. What happens when the power fails? Every serious pilot should own a handheld transceiver. Navigation capability is nice, but not essential...a controller can usually locate your plane on radar and give you vectors to a safe harbor.

What if the ATC facility is off your right wing, and you are sitting in the left seat? That "rubber ducky" antenna that came with the handheld won't be too efficient. If you own your own airplane, have an external antenna installed with a pigtail into the cockpit where it can be attached to the handheld. If you are a renter, take your handheld to an electronics shop and get a cable to connect your handheld to its rubber ducky, and figure out some way (suction cups, duct tape, whatever) to attach the rubber ducky to the window on the side toward ATC. Mount it vertically.

A handheld also comes in handy for monitoring aviation communications on the ground, just to stay sharp on how things are done.

A Matter of Procedure

Why You Have Two Ears and One Mouth

The most important rule of communication is this: Listen before you talk. Aviation radio is a huge party line, with dozens of people involved at any one time. If you switch frequencies and make an immediate transmission without listening first, your transmission may block communication from another airplane. It's not like a telephone, where both parties can talk at once and still hear one another. Several bad things happen when you fail to listen before transmitting: Your transmission may not be heard, another airplane may miss an important transmission because of your carelessness, or two simultaneous transmissions on the same frequency will cause a loud squeal in the headsets of everyone monitoring the frequency. Also, if you rush your transmission the first syllable or two will be cut off...so hesitate for a second before you begin to talk.

In normal, day-to-day conversation, if you heard one person ask another "Where do you work?" you wouldn't even consider interrupting with your own question until the first person's question had been answered. You know from experience that a question will elicit a reply, and that to interfere is bad manners. In the world of aviation communication, a transmission may not sound like a question but you should still wait for a reply.

> **PILOT** *"Bigtown Tower, Piper 70497 six miles south for landing with Echo"*

—doesn't sound like a question, but you know that the tower is going to give 497 some form of instruction or clearance and that you should not key your mike until the exchange is complete. Always listen for a few seconds before you transmit.

Is this a mistake you could make inadvertently? Unfortunately, it is—and there is nothing you can do about it. The FAA's communication network includes remote transmitter/receivers, and it is possible for a pilot too far away for you to hear to be calling an FAA facility on a remote frequency that you plan to use. If the controller says,

TOWER *"Piper 70497 stand by, you stepped on another transmission,"*

just keep quiet—if the controller can hear both pilots but you can't hear one another, there's nothing for you to feel bad about.

Similarly, you may listen on a frequency for a few seconds, hear nothing going on, and have your call to an FAA facility answered with,

TOWER *"Piper 70497, stand by, I'm working other traffic."*

In this situation, the controller might be talking to a military flight on an ultra-high aviation frequency (UHF) that your radio cannot receive. Those are the breaks of the game.

Never feel that you are doing an ATC facility a favor by not bothering the controllers with a call. The traffic count (number of communications) can affect staffing levels, pay scales, and even the continued existence of a facility.

What's In A Name?

Each type of ATC facility is addressed in a specific way. When you call a Flight Service Station, use **Radio**:

> **PILOT** *"Norfolk Radio, Baron 1014W 15 miles north, request traffic advisory."*

A call to a control tower uses **Tower**:

> **PILOT** *"Portland Tower, Baron 1014W ready for takeoff."*

On the ground at a tower-controlled airport, use **Ground** when calling the ground controller:

> **PILOT** *"Charlotte Ground, Baron 1014W off runway five, taxi to the ramp."*

If you are calling a radar controller at an approach/departure facility after being told by a Center or tower controller to contact that facility, just follow instructions:

> **CONTROLLER** *"Baron 1014W contact Bay Approach 135.4"*

—which should be acknowledged with,

> **PILOT** *"14W going to approach,"*

and after listening on 135.4 for a few seconds,

> **PILOT** *"Bay APPROACH, Baron 1014W level at 2,500, VFR."*

If you are told to call **Departure** Control, of course, that is who you would call. But what if you want to make an initial contact with a radar facility and don't know whether to address it as approach or departure—use **Approach**.

At many airports, airspace sectorization means that when the wind shifts and the active runway changes, a given location might change from approach's airspace to that owned by the departure controller. This kind of thing is just shrugged off by the controller, so don't worry about who to address—they will respond regardless of which name you use.

You will call an air route traffic control center when you want radar flight following on a cross-country trip, and you should address that facility as **Center**:

> **PILOT** *"Fort Worth Center, Piper 70497 over Blue Ridge VOR, request VFR flight following to Waco."*

Center frequencies can be found in the Airport/Facility Directory.

A UNICOM station would, of course, be addressed as **UNICOM**.

But what if the name of the airport is "General William J. Fox Airfield," "Charles B. Wheeler Downtown," or "Central Illinois Regional at Bloomington-Normal"? Best thing to do is listen to the tower or UNICOM frequency to see what others are using, or what the tower/UNICOM itself is using. Second best is to look in the A/FD: If the tower frequency is listed as "Downtown Tower," you have it made; if not, use the name of the town until corrected...it's not a big deal.

When you need weather information or want to make a pilot report, use **Flight Watch** on 122.0 MHz (this is only monitored from 0600 until 2200 local time). You will find locations of Flight Watch stations and their remotes on the inside of the back cover of the A/FD.

> **PILOT** *"Houston Flight Watch, Baron 1014W over Lufkin, VFR to Waco, can I have the latest Waco weather?"*

Controllers have heard everything, so if you use the wrong name by mistake the controller will usually reply using the proper name without missing a beat. *See* Appendix A for a comprehensive discussion of how to address each facility.

According to the AIM, you are not supposed to shorten your callsign to the last three numbers/letters (14Whiskey instead of 1014Whiskey) until the controller replies using the shortened callsign. They need the complete callsign at least once.

Don't talk to another airplane on an ATC frequency…that's what air-to-air is for. And don't use 123.45; that frequency is assigned to aircraft manufacturers and is also used for trans-oceanic communications. But what if you hear another airplane call the tower and report a position very close to yours, and you do not have visual contact with that airplane? Ask the tower controller to ask the other airplane for its exact position and altitude…the other pilot will hear your request and will probably volunteer the information you need.

The same "party line" philosophy works when you hear another pilot report turbulence, icing, or anything else that might affect your flight. Ask the controller to ask the other pilot for details, and you will get what you need.

Say It All in One Breath

When transmitting to ATC, don't add last minute thoughts. For example, if you say:

PILOT *"Centennial Tower, Baron 1014W ready for takeoff…we're VFR to Grand Junction,"*

there is an excellent chance that the controller will key the mike and begin to reply as soon as you say "takeoff" and pause. The result will be a squeal on the frequency as you and the controller transmit simultaneously, and the controller will miss your comment about your destination.

"Roger, Wilco, Over and Out"

This phrase is often incorrectly used as an example of communication overkill. There is no question that with the sophistication of today's radios, "over" and "out" are pretty much outdated. However, the terms **Roger** and **Wilco** still play an important part in communications with ATC.

You would never know it from listening to your VHF radio, but Roger is a term that should be used very infrequently. It simply means "I have received your transmission," and nothing more. It does not mean "yes" or "I will comply." For example, if the tower controller says,

TOWER *"Piper 70497, you are number two for takeoff,"*

you can reasonably reply "Roger," because it would be foolish to say "yes" or "I will comply." If a controller asks if you are over the blue water tower, however, your answer should be "affirmative," not "Roger."

The great majority of ATC transmissions should be answered with Wilco which means "I understand and will comply":

TOWER *"Piper 70497, follow the Piper on downwind"*

PILOT *"Wilco, 497"*

TOWER *"Piper 70497, contact Seattle Approach Control on 119.2"*

PILOT *"119.2, Wilco, 497"*

TOWER *"Piper 70497, taxi into position and hold"*

PILOT *"Wilco, 497"*

There is one situation in which Wilco won't do the job. If you are directed to hold short of a runway you must confirm this instruction by repeating the directive:

PILOT *"70497 holding short of runway 21."*

A controller at Chicago O'Hare tells a story about a pilot who called for permission to cross runway 14. "Hold short," was the reply. The pilot immediately added power and taxied across the active runway. "Why did you cross the runway?" he asked the errant pilot. "I asked for permission and you said 'Oh, sure'..." the pilot replied.

Did you notice that whenever I give an example of a reply I put the callsign last? The AIM says that the callsign should be first, so why do I recommend the opposite? Because the controllers prefer that you transmit your callsign last.

Denny Cunningham, a tower controller at O'Hare Airport in Chicago, the world's busiest airport 51 weeks out of the year (Wittman Field in Oshkosh, Wisconsin, gets that honor during the week of the Experimental Aircraft Association's annual fly-in), frequently addresses pilot groups on communications subjects. He says that although two pilots might key their microphones at the same time, it is unlikely that they will both finish talking at the same time. With the callsign transmitted last, ATC is in a better position to clear up misunderstandings.

Other Readbacks

Although you will seldom receive vectors or altitude changes from ATC while operating in VFR conditions (this will most likely happen in Class B airspace or when being vectored in Class C airspace), changes in heading or altitude should be read back:

> **CONTROLLER** *"Piper 70497, climb and maintain 4,000"*
>
> **PILOT** *"Leaving 2,000 climbing 4,000, 497."*
>
> **CONTROLLER** *"Piper 70497, turn right heading 330"*
>
> **PILOT** *"Right heading 330, 497."*

You should acknowledge transmissions from ATC if for no better reason than to let the controller know that you received the transmissions. At very busy airports, however, the tower controller may be giving instructions to several airplanes at once without pausing for breath. You know that you will cause interference if you key your microphone while the controller is still transmitting, so what do you do? The answer has to be to do nothing. Follow the instructions that apply to your flight without acknowledgment.

Although I write "2,000" in this book, in real life you say "two thousand." Readback of altitude changes should include the word "thousand." This is not excess verbiage—it is necessary.

Be Brief...But Clear

To a controller, time is like money...not to be wasted. Every second you save when transmitting is a second that the controller can use to talk to someone else, and those saved seconds add up over a controller's shift.

As much as possible, avoid prepositions (to, for, with), articles (a, an, the), conjunctions (and, but), adjectives, adverbs, and even verbs. Don't say, "taking the active" or "traffic in the area, please advise":

> **PILOT** *"Podunk Ground, Cessna 1357X is at the south ramp with information Echo, request taxi instructions to the active runway; I'm on a VFR flight to Bigburg."*

Instead,

> **PILOT** *"Ground, Cessna 1357X, south ramp, Echo, VFR Bigburg."*

Never make brevity more important than clarity, however. Use plain language whenever necessary to get your point across. And don't use Citizen's Band slang (ten-four) or fake military lingo (tally-ho, no joy).

The majority of controllers are not pilots. "We're in the soup" means nothing to a controller; a fatal accident occurred because a controller thought it meant no more than that visibility was poor.

One of the most useless phrases in aviation communications is "Request permission to…" You don't have to request permission to do anything. Just make your request:

PILOT *"Request taxi instructions."*

PILOT *"Request higher/lower."*

PILOT *"Request crosswind departure."*

The word "permission" is not necessary.

"Say Again?"

When in doubt, never hesitate to say "Say again?" It is far better to eliminate any doubt than to take action on what you think you were directed to do, and it is easier on the controller's nerves as well.

Similar Callsigns

Although it happens infrequently, you may find yourself flying in the same patch of air as another airplane with a similar callsign. This can be more than an inconvenience if the pilot of one airplane acknowledges or follows instructions meant for the other pilot. Controllers are pretty sharp at picking up on this possibility and will warn both pilots:

CONTROLLER *"Piper 234D, be aware that Piper 334D is on the frequency."*

However, you should not rely on the controller but should be alert to the possibility of confusion if you hear an airplane on the frequency with a callsign close to yours.

Don't hesitate to speak up if you hear an error on the frequency. You may cause some embarrassment, but that is better than bent metal:

PILOT *"Center, Piper 70497—I believe that Piper 1234G acknowledged the heading change you gave Piper 1334G."*

Controllers are human—they don't catch all of the readback errors.

Another line of defense is situational awareness—if you think that an ATC instruction is addressed to you but it doesn't make sense, speak up. If you are flying Piper 234D and have just reported to the tower that you are five miles out on the

45, an instruction to Piper 234D to "turn base and follow the Baron on final" is obviously not directed to you. You can handle this by saying either "Was that for Piper 234D?" or "Piper 234D is still on the 45."

Type Confusion

Controllers occasionally confuse airplane types, and you must be ready to politely correct the controller when this happens. Controllers make traffic flow decisions based on the performance of the model they think you are flying (they have a quick reference sheet with models and speeds). If you are flying Arrow 369OJ and the controller keeps calling you Cessna 369OJ, it is your responsibility to get the situation straightened out. Plain language works best:

PILOT *"Tampa Approach, 369OJ is a Piper Arrow, not a Cessna."*

Bear in mind that aircraft registration numbers are unique—the same numbers cannot be assigned to two different aircraft even if they are different models or types. We like to keep airborne communications brief and to the point, but if you don't make it clear to the controller just what kind of airplane you are flying, you are not doing your part:

PILOT *"Atlanta Center, 8452B is a SINGLE-ENGINE Aero Commander."*

You should not use the prefix "November" as part of your initial call to an ATC facility. The tail numbers of all U.S.-registered airplanes begin with N, so when you say November you are not telling the controller anything useful.

You should use your full callsign when making the initial call to a station:

PILOT *"Norwood Tower, Piper 70497 six miles east for landing with Oscar."*

Once communications have been established, you are allowed to use the last three letters or digits. Whenever there is a possibility of confusion with another airplane, use the full callsign preceded by the aircraft type. Confusion breeds disaster.

As you monitor aviation frequencies, you may hear some unfamiliar identifications. **Lifeguard**, followed by an aircraft identification, indicates an air ambulance or emergency helicopter. You should give this airplane's communications priority over yours (it will get priority handling by ATC anyway). **Tango**, used after an aircraft identification, is supposed to be used by air taxi operators. Very few do so, but you may hear it on occasion. An FAA flight test airplane will identify itself as **Flight Test**, and this is a signal for a high level of caution on your part—they fly as low as 50 feet above the runway and frequently fly in the face of normal traffic. If you hear **Air Force One** (or **AF Two**), don't be surprised if area airports are closed for awhile. The Secret Service likes to sterilize the airspace when the

President or Vice President are in the area. **Experimental** could mean anything from a KitFox to a Bede jet; keep your eyes open because you can't make any assumptions.

Don't forget the words **Unable** and **Unfamiliar** when communicating with ATC. They are the words to use when you are asked to do something unsafe, or when you are in an area where you are a stranger.

Communicating an Emergency

The emergency VHF frequency is 121.5 MHz, and if you have a problem and don't have time to look up a Center frequency you should switch to 121.5 (you are supposed to monitor 121.5 MHz at all times, if possible). Following international procedures, the AIM says that if you are in an urgency situation, rather than a distress situation, you should start your transmission on the emergency frequency by saying "Pan, Pan, Pan," followed by your airplane's identification. An urgency situation might be disorientation, while a distress might be empty fuel tanks. In over thirty years of flying I have never heard anyone say "Pan, Pan, Pan"; I have heard some "Mayday" calls, however. Mayday is the internationally-accepted word for distress, and it gets the attention of everyone on the frequency. Switch to 121.5 and say:

PILOT *"Mayday, Mayday, Mayday, Piper 40497 calling any station."*

That should result in more than one reply—choose the strongest signal and report your situation to that facility.

Your emergency locator transmitter broadcasts on 121.5, but that will change to 406 MHz; 121.5 is being phased out. The new 406 MHz beacons transmit your aircraft tail number and (some models) your GPS-derived position.

Even though you are not talking on or monitoring any ATC frequencies, your transponder should be ON and set to 1200, the VFR code, at all times. The transponder return from your airplane will show up on the controller's scope as an anonymous "1200 squawk"; the controller will have no idea of who you are or at what altitude you are flying, but when you switch to 7700 your return will get immediate attention. ATC records and stores the information displayed on radar scopes, and even if you don't say a word to anyone or squawk 7700 it is possible for them to track your 1200 squawk until it disappears. By the time this takes place, however, you will have been missing for several hours and your survival skills will have been put to the test. This is nice information to know, but you have so many communication options available that your survival should never depend on this procedure. If you think you are lost or if the engine is making funny noises, tell someone about it right away.

If you are receiving radar flight following service and the controller asks you to intensify your transponder return by pushing the "ident" button, accompany that action with a verbal confirmation:

> **PILOT** *"Cessna 1357X, ident."*

Many instructors counsel against this, reasoning that the change in your return on the controller's scope is confirmation enough. Ask any controller about this, however, and you will be told that because of the rate at which the radar sweeps and the screen is refreshed, it may take a couple of sweeps before your enhanced return is seen. In the meantime, the controller has asked you to do something and cannot be sure that you received the transmission. So speak up.

Simplex vs. Duplex

Aviation communications can be characterized as either simplex or duplex. When you and the controller transmit and receive on the same frequency, which is the usual situation, it is a simplex channel. Put simply, you could not both speak at once and be able to hear one another. There are a few remaining situations in which you will use duplex, where you transmit to ATC on one frequency and listen on another. The most common duplex situation has you transmitting on 122.1 MHz and listening on a VOR frequency; the controller listens to you on 122.1 and transmits on the VOR frequency.

Duplex is being phased out across the country but still exists in some locations. If you look at a VOR frequency box on a sectional chart and see "122.1R" printed above the box it means that a Flight Service Station specialist is listening to a speaker connected to a receiver tuned to 122.1 MHz and will hear your call. That specialist has just one microphone and a selector switch that permits the mike to be connected to one of the many VORs or Remote Communication Outlets in that FSS's area. If you don't see the R, it is a simplex frequency and listening on the VOR won't work.

When a call is received, a light blinks on the specialist's panel if the calling pilot is flying at an altitude that allows reception at only one remote receiver site. If the pilot is flying at an altitude high enough so that the signal reaches two receivers on the same frequency, two lights blink and the specialist has to guess which remote transmitter to use when replying. The solution is simple—a pilot using duplex should always include the name of the VOR in the transmission:

PILOT *"Macon Radio, Baron 1014W, listening to the Dublin VOR."*

The practice of including the name of the VOR in your initial call is equally applicable to the Enroute Flight Advisory Service, or Flight Watch. When calling for weather information on 122.0 MHz, say:

PILOT *"Gainesville Flight Watch, Piper 70497, over the Jacksonville VOR."*

I Love A Mystery

Let's say that you are calling Cedar City Flight Watch to check the weather at your destination:

PILOT *"Cedar City Flight Watch, Baron 1014W over Malad,"*

and the response is:

FLIGHT WATCH *"Aircraft over Malad, say again callsign."*

This simply means that the Flight Watch specialist was busy doing something else and did not catch your callsign. Just provide the information and go on with your request—communications have been established:

PILOT *"Cedar City, Baron 1014W over Malad, request current Rock Springs Weather."*

A variation of the same scenario is:

FLIGHT WATCH *"Aircraft calling Cedar City Flight Watch, say again callsign."*

Your transmission has been heard by the ground station, but the specialist is not sure that he or she has logged the callsign correctly.

Hooked On Phonics

Use the phonetic alphabet as published in the *Aeronautical Information Manual*. It may bring a smile to the faces of pilots on the frequency when they hear ATIS information Whiskey referred to as the "Booze News," or King Air 76MM calling in as "76 Mickey Mouse," but it sets a poor example for students and is incomprehensible to pilots for whom English is a second language. The ICAO phonetic alphabet was developed so that pilots from all nations could understand one another.

Also, communicating numbers with accuracy is critically important in aviation. The AIM tells you how to pronounce individual digits to avoid confusion. "Five" and "nine" are the digits most frequently misinterpreted, and you will soon be saying "niner" like a pro.

The AIM tells you to say "four thousand, five hundred" for 4,500. You will hear many pilots at all levels in the aviation industry say "four point five" when reporting that altitude. I cannot argue with doing it the AIM way, but I have no problem with using the decimal notation because in my opinion it is not confusing to either the controller or other pilots on the frequency.

A recent change to the *Air Traffic Control Handbook* has created the best of both worlds: Controllers can now say,

> **CONTROLLER** *"Baron 1014W, descend and maintain one-zero-thousand, that's ten thousand."*

To gain the most benefit from this change, pilots should begin this practice. (*See* the AIM and ICAO table for telephony and phonetic alphabet on the next page.)

Character	Morse Code	Telephony	Phonic (Pronunciation)
A	● —	Alpha	(AL-FAH)
B	— ● ● ●	Bravo	(BRAH-VOH)
C	— ● — ●	Charlie	(CHAR-LEE) or (SHAR-LEE)
D	— ● ●	Delta	(DELL-TAH)
E	●	Echo	(ECK-OH)
F	● ● — ●	Foxtrot	(FOKS-TROT)
G	— — ●	Golf	(GOLF)
H	● ● ● ●	Hotel	(HOH-TEL)
I	● ●	India	(IN-DEE-AH)
J	● — — —	Juliett	(JEW-LEE-ETT)
K	— ● —	Kilo	(KEY-LOH)
L	● — ● ●	Lima	(LEE-MAH)
M	— —	Mike	(MIKE)
N	— ●	November	(NO-VEM-BER)
O	— — —	Oscar	(OSS-CAH)
P	● — — ●	Papa	(PAH-PAH)
Q	— — ● —	Quebec	(KEH-BECK)
R	● — ●	Romeo	(ROW-ME-OH)
S	● ● ●	Sierra	(SEE-AIR-RAH)
T	—	Tango	(TANG-GO)
U	● ● —	Uniform	(YOU-NEE-FORM) or (OO-NEE-FORM)
V	● ● ● —	Victor	(VIK-TAH)
W	● — —	Whiskey	(WISS-KEY)
X	— ● ● —	Xray	(ECKS-RAY)
Y	— ● — —	Yankee	(YANG-KEY)
Z	— — ● ●	Zulu	(ZOO-LOO)
1	● — — — —	One	(WUN)
2	● ● — — —	Two	(TOO)
3	● ● ● — —	Three	(TREE)
4	● ● ● ● —	Four	(FOW-ER)
5	● ● ● ● ●	Five	(FIFE)
6	— ● ● ● ●	Six	(SIX)
7	— — ● ● ●	Seven	(SEV-EN)
8	— — — ● ●	Eight	(AIT)
9	— — — — ●	Nine	(NIN-ER)
0	— — — — —	Zero	(ZEE-RO)

Figure 3-1. AIM: telephony and ICAO phonetic alphabet

There's Not Much Of It

According to the most recent statistics, there are about 17,000 airports in the United States and less than 700 of them have control towers. That's why our discussion of communications will literally start at the grassroots level and work toward the complex airspace at busy airports with airline traffic. Although I will refer to and illustrate actual airports, you should understand that whatever works in these situations will apply to any similar airport.

With very few exceptions, Class G airspace exists from ground level up to either 700 feet or 1,200 feet above the ground. There are a few locations (in the western United States) where it extends all the way to 14,500 feet, but for the purposes of this text I am going to ignore them.

Figure 4-1 is an aerial view of the Sky Harbor airport in Sultan, Washington (airport A on the four-color sectional chart foldout at the back of the book). It is a typical small airport with a couple of rows of hangars and direct access to the turf runway. According to the sectional, no services are available because the airport symbol doesn't have little protrusions sticking out of it. The fact O that the symbol is not solid magenta indicates that the runway surface is not paved.

Your sectional is speckled with similar airport symbols; some have an R for private or non-public airports. These are all perfectly good airports whose owners have removed them from public use status due to liability concerns.

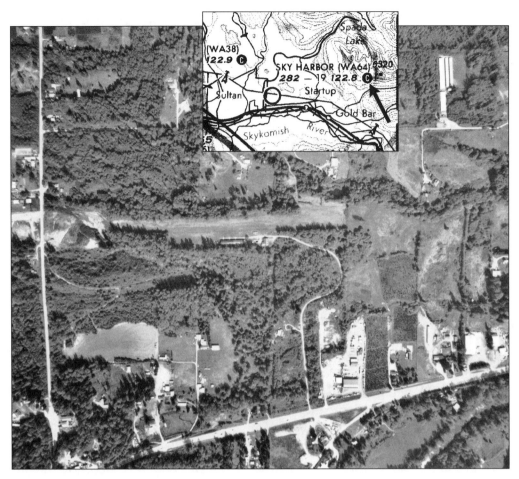

Figure 4-1. Sky Harbor Airport, Sultan, Washington

Sky Harbor does not have magenta tint surrounding it, and the boundary of air-space with a 700-foot AGL floor is just west of the airport, so you must assume that it lies in Class G airspace and that the floor of the overlying Class E airspace is at 1,200 feet AGL. You can taxi out and take off from an airport like Sky Harbor without saying a word to anyone, although the airport does have a UNICOM station as indicated by the frequency 122.8, and the letter C in a magenta oval meaning that 122.8 is the Common Traffic Advisory Frequency (CTAF). It is around airports like Sky Harbor that most airplanes without radios or electrical systems (NORDO—"no radio") can be found.

Does this mean that you should not use your radio if you have one? No way. Use your radio as a tool to increase the safety of flight by staying alert to the positions and intentions of other radio-equipped planes, while keeping a wary eye out for NORDOs.

UNICOM and MULTICOM

A UNICOM is a ground station purchased by the airport operator, licensed by the FCC, and operated by employees of the FBO. A UNICOM has no official standing whatsoever. The *Aeronautical Information Manual* says that a UNICOM is an "aeronautical advisory station," whereas a flight service station is an "airport advisory station." Although UNICOM and MULTICOM look like acronyms, the letters do not mean anything.

Any information you receive from a UNICOM is wholly advisory in nature and should be supplemented by your own observations. If the UNICOM operator says there is no traffic in the pattern and you almost get T-boned on downwind, that is not the fault of the UNICOM operator. If the UNICOM operator says runway 9 is in use and you want to take off on runway 27, have at it (but you will probably get some heat from other pilots in the area who are on the frequency).

UNICOM frequencies are shown on aeronautical charts, of course (for example, **122.700 122.725 122.800 122.975 123.00 123.050 123.075**).

Some Fixed Base Operators have their own discrete frequencies, not shown on any chart or included in any FAA publication. They would be useful for questions about parking, fueling, etc. You will find them in places like www.airnav.com,

Flight Guide, or by calling the FBO at your destination before takeoff (always a good idea). This is neither UNICOM nor MULTICOM, so your call would be addressed to the name of the operator: "BirdRoost Aviation, Piper 70497, where can I park for an overnight stay?"

The most common use of the UNICOM frequency is obtaining information on the runway in use. However, you can use it to ask the operator to call a taxi, to call someone and advise of your estimated time of arrival, to ask for a quick turnaround from the fuelers, to request two hamburgers, one with onions, one without, and two chocolate milkshakes be ready when you land. In other words, it is good for just about anything within reason except official weather reports or takeoff and landing clearances.

You will know that the airport has a SuperUnicom® if a computer-generated human voice answers your call. SuperUnicoms are automatic devices that can supply temperature, wind direction and velocity, altimeter setting, and runway in use (based only on airplanes in the pattern that have called in). Click your mike button three times to activate the device.

MULTICOM, 122.9 MHz, is the Common Traffic Advisory Frequency at airports that have no UNICOM. Accordingly, when you use 122.9, your transmissions are to other airplanes in flight and not to anyone on the ground.

Let's Go Bore Some Holes in the Sky

Having said all that, let's take off from Sky Harbor. It is like the majority of small airports in having only one runway and a broad area that leads from the ramp/tie-down area to the runway, although the path from the ramp to the runway is usually better defined. There is no taxiway paralleling the runway. This means that departing airplanes must taxi in the face of arriving traffic to get to the runup position (back-taxiing), and that pilots who roll past the midfield exit after landing must do a 180 and taxi back to the exit.

As noted before, Sky Harbor's Common Traffic Advisory Frequency is 122.8 MHz; other UNICOM frequencies are 122.7 and 123.0, with some others being reserved by the Federal Communication Commission for UNICOM assignment when necessary. If Sky Harbor had no ground radio station at all, the CTAF would be 122.9, which is the MULTICOM frequency.

AIRPORT OPERATIONS

Single Runway

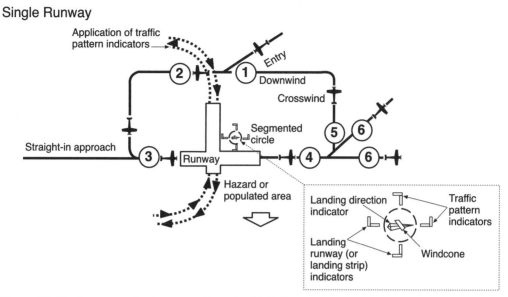

Figure 4-2. Traffic pattern, single runway (from AC 90-66A)

While still on the taxiway at Sky Harbor before entering the runway, you transmit on 122.8 MHz, saying:

PILOT *"Sky Harbor traffic, Piper 70497 back-taxiing for takeoff on runway 27, Sky Harbor."*

Why don't you address your transmission to Sky Harbor UNICOM? Because your intentions to back-taxi are of no interest to the person in the airport office, only to the pilots operating in the vicinity of the airport. By the way, never say, "Taking the active" — always include a runway number. At an uncontrolled airport, all runways can be used.

Before you actually taxi onto the runway, it is a good idea to do a 360-degree turn, or maneuver as necessary, to see the final approach from both directions and to check for airplanes on base as well. Sky Harbor has no pattern direction markings, according to the photograph, so a standard left-hand traffic pattern is in effect. (*See* Figure 4-2.) When you are checking for airplanes on base leg, however, don't forget that all pilots are not as smart as you are and someone might very well be on right base. And don't forget about the NORDO (no radio) airplanes that couldn't hear your announcement in the first place.

While taxiing east toward the runup area for runway 27, monitor 122.8 so that if someone makes an announcement about entering the pattern (or worse, reports 3 miles out on final) you can repeat that you are back-taxiing and will clear the runway at the end.

Most airports of this type have provided a wide spot at each end of the runway so pilots can perform their pretakeoff checks clear of the runway surface. If this is not the case at an airport you are using, complete your runup on the taxiway or ramp before taxiing onto the runway (use good etiquette—turn so that your propwash won't affect airplanes taxiing behind you).

If you are of a slow and methodical nature—and that's better than being rash and impetuous—then perform as much of the runup as possible before you even enter the taxiway, wherever one exists. This also applies to instructional flights where the instructor explains at great length every detail of the runup procedure. In other words, don't block the taxiway any longer than absolutely necessary.

When you have completed your pretakeoff preparations and are ready to go, take a good long look at the base legs and final approach again and then transmit:

> **PILOT** *"Piper 70497 departing Sky Harbor runway 27, standard departure"*
> *[or "straight-out," or "climbing," or "downwind departure"].*

Either "staying in the pattern" or "closed traffic" is appropriate if you are just going to practice takeoffs and landings at Sky Harbor. In that situation, saying,

> **PILOT** *"Piper 70497 is staying in the pattern at Sky Harbor."*

—serves to advise everyone in the area who is monitoring the frequency.

The idea is to give approaching pilots some insight into where you are going so they can avoid you.

 For the purpose of this exercise, however, you are going to fly about 13 miles to Harvey Field in Snohomish (airport B on the sectional chart excerpt) for one of those $100 hamburgers that pilots enjoy so much. If you remain less than 1,200 feet above the ground until you reach the edge of the magenta tint (see the glider symbol?), then duck down to less than 700 feet above the ground, you will stay in Class G airspace. During daylight hours, you will need to avoid touching a cloud with any portion of your airplane and have one mile flight visibility, if you are to fly legally under basic visual flight rules. Flying in those conditions is not the smartest thing to do, however, because you will be bending (if not breaking) the minimum altitude rules, and the hamburger almost certainly will not be worth the risk of violating the Federal Aviation Regulations.

Figure 4-3. Harvey Field, Snohomish, Washington

To add a little fun to the trip, your buddy Joe is going to fly his Bellanca to Harvey Field (Figure 4-3) so that you can commiserate about the cost of hamburgers (and annual inspections). After you have both departed from Sky Harbor you can get together on the plane-to-plane frequency, 122.75 MHz, and chat about the scenery. This frequency is the only frequency approved for plane-to-plane chatter between airborne pilots. Note that 123.45 is *not* an approved frequency, no matter what other pilots and instructors tell you. Do not use the CTAF for air-to-air communication.

By the way, if someone tells you that 123.45 MHz is an acceptable plane-to-plane frequency, keep one hand on your money. The Federal Communications Commission guards its frequency assignment duties jealously and will fine pilots who misuse their radios. Out here in the Pacific Northwest, 123.45 MHz is assigned to the Boeing Company for flight-test purposes; someone else probably has been assigned that frequency in your area. You can be sure that it has not been allocated to anyone for plane-to-plane communication.

You Want Fries With That?

The airport symbol for Harvey Field is solid magenta and shows one runway, which means it has a hard-surfaced runway more than 1,500 feet long, and it has the little protrusions that mean "services" to the mapmakers and "fuel" to pilots; CTAF is 123.0 MHz. Before you leave the Harvey Field area on the sectional chart excerpt, note the parachute symbol just west of the field. It might be a good idea to check in with the UNICOM and check on jumping activity:

> **PILOT** *"Harvey UNICOM, Piper 70497 ten miles east, landing Harvey. Could you give me the runway in use and advise of any parachuting activity?"*

> **UNICOM** *"Piper 70497, Harvey UNICOM, runway 32 is in use, two aircraft in the pattern, no jump activity."*

> **PILOT** *"70497, understand runway 32."*

Note from Figure 4-3 that there is a pattern indicator with a tetrahedron just east of runway 32. It is indicating that the traffic pattern is to the west; that is, left traffic for runway 32 and right traffic for runway 14. This type of restriction is common when the local municipality wants to keep airplanes from flying over the city. At Harvey Field, the downwind leg is flown over the fields west of the airport no matter which way the wind is blowing.

If you are a fairly inexperienced pilot who is being trained at a tower-controlled airport, you must understand that UNICOM is diametrically opposite from tower control. The UNICOM transmitter/receiver is supposed to be on during operating hours, but that doesn't mean that there is someone standing close by just waiting for your call. The UNICOM operator might be out on the ramp pumping gas, calling a motel for transportation, or just out of the office for a few minutes. In other words, you should not be surprised if you call UNICOM and fail to get a reply.

Switch to the UNICOM frequency about 10 miles out. Devote a few minutes to just "reading the mail" (that's ham operator talk for "just listening in"). Chances are that you will hear other airplanes in the Harvey Field area getting runway information from the UNICOM operator and you can use that information to plan your approach to the pattern. If you hear,

> **PILOT** *"Harvey UNICOM, Piper 70497, airport advisory."*

> **UNICOM** *"Piper 70497, Harvey UNICOM, Runway 32 is the active; there are three in the pattern."*

—you have the information you need to plan an entry to the traffic pattern. If there isn't anyone on the frequency who is flying near Harvey you will probably hear transmissions from pilots flying near other airports that share 123.0 MHz with Harvey Field. The UNICOM frequencies can become really crowded on a nice day, and you have to listen closely to pick out the transmissions that apply to your destination airport.

The *Aeronautical Information Manual* suggests that pilots "self-announce," their position when about 10 miles from the destination airport, giving their identification, position in relation to the airport, altitude, and intentions:

> **PILOT** *"Harvey traffic, Piper 70497 10 miles east, two thousand five hundred feet, landing [or 'touch and go'] Harvey."*

It is important to mention the name of the airport at both the beginning and end of each self-announced transmission so that pilots monitoring the frequency will know at which airport you are operating.

The Tower of Babble

The AIM also suggests that pilots operating at non-tower airports make self-announced reports on downwind, base, final and when exiting the runway:

> **PILOT** *"Harvey traffic, Piper 70497 entering downwind, Runway 32 Harvey."*

> **PILOT** *"Harvey traffic, Piper 70497 turning left base, Runway 32 Harvey."*

> **PILOT** *"Harvey traffic, Piper 70497 on final, touch and go, Runway 32 Harvey."*

> **PILOT** *"Harvey traffic, Piper 70497, clear of Runway 32."*

A little bit of common sense helps a lot in this situation. I yield to no one in my respect for the regulations and approved procedures, but if every pilot made every report suggested by the AIM at every small airport on a sunny weekend afternoon the UNICOM frequencies would be saturated. After a few minutes in the traffic pattern you will have a handle on just how much activity there is and how necessary it is for you to make transmissions that no one is listening to. Certainly, if there are four airplanes in the touch-and-go pattern and itinerant airplanes are coming in to make full-stop landings, safety dictates position and intention reports. On the other hand, if you have the pattern to yourself you might want to reconsider making *every* report suggested by the AIM.

By the way, your airplane will be more visible to other pilots when its wings are banked, so I like "turning downwind," "turning base," "turning final," instead of reporting when established on one of those legs of the pattern.

Flying from Sky Harbor to enter Harvey's pattern west of the field means flying over the airport. Commonly accepted practice is to overfly the airport 500 feet above pattern altitude, fly beyond the downwind leg giving yourself room to get turned around and, after descending to pattern altitude, entering the downwind on the 45 at midfield. (And in this case, don't fly into Paine Field's airspace while you are getting turned around.)

When is a Tower Not a Tower?

Is the airspace around an airport with a tower ever Class G airspace? Can you tell by looking at the sectional chart? The answers are yes and no. There are tower-controlled airports (blue airport symbol) with a magenta-tint circle indicating that the floor of controlled airspace is at 700 feet above the ground. Very few of them, but they do exist and they are to be treated like any towered airport (*see* Chapter 6). When their controllers shut down for the night, the airspace remains Class G. But at a "normal" towered airport, when the last controller shuts off the coffeepot and turns out the lights, the airspace can become either Class G or Class E (Chapter 5). The only place you can look to find out which is the Airport/ Facility Directory.

If there will be a human weather observer (or an Automated Weather Observing Station/Automated Surface Observing Station) there all night, then it is Class E airspace all the way to the ground and you must get a Special VFR clearance (*see* Chapter 5) if you want to depart in weather that is less than VFR. If you are not instrument rated, you can't legally ask for a SVFR clearance at night. One mile visibility and clear of clouds is necessary to operate in Class E surface area under SVFR.

If no weather observer will be present, then the airspace from field elevation to 700 feet above the ground becomes Class G airspace and once again you need only one mile visibility and must remain clear of clouds during daylight hours... three miles visibility is required at night.

On the sectional chart excerpt, note the Olympia Airport (C) and the Tacoma Narrows Airport (D). Both have part-time towers, as indicated by the asterisk next to the tower frequency. Both are surrounded by blue dashed lines, indicating the horizontal dimensions of the tower's class D airspace when the tower is in operation. Now look at Figure 4-4, containing a portion of the Airport/Facility Directory listing for both airports: Olympia's airspace becomes Class E when the controller goes home, but because Tacoma Narrows has no full-time weather observer, its airspace becomes Class G when the controller goes home. You just can't rely on the sectional chart to tell you everything you need to know. While you are looking at Figure 4-4, check out the Communications listings; that's where you find the frequencies you need.

- -

TACOMA NARROWS   (TIW)   4 W   UTC–8(–7DT)   N47°16.08' W122°34.69'   SEATTLE
   292   B   S4   FUEL 80, 100, 100LL, JET A   OX 4   TPA–1300(1008)   LRA   H–1A, L–1D
   RWY 17-35: H5002X150 (ASPH–AFSC)   S–50, D–80, DT–80   MIRL   IAP
      RWY 17: MALSR. PAPI(P4L)—GA 3.0°. TCH 50'. Trees. Rgt. tfc.   RWY 35: REIL. VASI(V4L)—GA 3.0° TCH 51'.
   AIRPORT REMARKS: Attended 1600–0400Z‡. Deer on and in vicinity of arpt. Noise sensitive arpt, for noise abatement
      and tfc procedures call arpt manager 206–591–5759 or 206–851–3544. Rwy 35 REIL out of svc indefinitely.
      ACTIVATE MALSR Rwy 17 and PAPI Rwy 17—CTAF. Landing fee acft over 12,500 lbs.
   WEATHER DATA SOURCES: LAWRS
   COMMUNICATIONS: CTAF 118.5   ATIS 124.05 (1600–0400Z‡)   UNICOM 122.95
      SEATTLE FSS (SEA) TF 1–800–WX–BRIEF. NOTAM FILE TIW.
   ® SEATTLE APP/DEP CON 120.1
      TOWER 118.5 (1600–0400Z‡)   GND CON 121.8
   → AIRSPACE: CLASS D svc effective 1600–0400Z‡ other times CLASS G.
   RADIO AIDS TO NAVIGATION: NOTAM FILE TCM.
      McCHORD (T) VORTAC 109.6   TCM   Chan 33   N47°08.86' W122°28.50'   308° 8.4 NM to fld. 285/22E.
      GRAVE NDB (MHW/LOM) 216   GR   N47°09.02' W122°36.29'   349° 7.2 NM to fld. NOTAM FILE GRF.
      ILS 109.1   I–TIW   Rwy 17   ILS unmonitored when twr clsd.
      COMM/NAVAID REMARKS: Emerg frequency 121.5 not avbl at twr.

OLYMPIA
   AERO PLAZA   (WA44)   5 SE   UTC–8(–7DT)   N46°59.56' W122°49.66'   SEATTLE
   213
   RWY 07-25: 2015X116 (TURF)
      RWY 07: Trees.   RWY 25: Trees.
   AIRPORT REMARKS: Unattended.
   COMMUNICATIONS: CTAF 122.9
      SEATTLE FSS (SEA) TF 1–800–WX–BRIEF. NOTAM FILE SEA.

- -

   OLYMPIA   (OLM)   4 S   UTC–8(–7DT)   N46°58.24' W122°54.17'   SEATTLE
   206   B   S4   FUEL 80, 100, 100LL, JET A   OX 1, 2   LRA   ARFF Index Ltd   H–1A, L–1C
   RWY 17-35: H5419X150 (ASPH)   S–55, D–69, DT–117   MIRL   IAP
      RWY 17: MALSR. Thld dsplcd 427'. Road.   RWY 35: VASI(V4L)—GA 3.0° TCH 50'. Trees. Rgt tfc.
   RWY 08-26: H5001X150 (ASPH)   S–30
      RWY 08: Ground. Rgt tfc.   RWY 26: Tree.
   AIRPORT REMARKS: Attended Apr–Sep 1600Z‡–sunset. For svc after hours call 206–754–4043 or 206–786–8333.
      Coyotes on and in vicinity of rwys. PPR for unscheduled air carrier ops with more than 30 passenger seats call
      arpt manager 206–586–6164. Rwy 08–26 and Taxiways F, G, C, E and D not avbl for air carrier ops with more
      than 30 passenger seats. NSTD hold short lgts intersection Twy E and Rwy 17. During hours twr clsd ACTIVATE
      MALSR Rwy 17 and VASI Rwy 35—CTAF. NOTE: See SPECIAL NOTICE—Land and Hold Short Operations.
   COMMUNICATIONS: CTAF 124.4   UNICOM 122.95
      SEATTLE FSS (SEA) TF 1–800–WX–BRIEF. NOTAM FILE OLM.
   ® SEATTLE APP/DEP CON 121.1
      TOWER 124.4 (1600–0400Z‡)   GND CON 121.6
   → AIRSPACE: CLASS D svc effective 1600–0400Z‡ other times CLASS E.
   RADIO AIDS TO NAVIGATION: NOTAM FILE OLM.
      (H) VORTACW 113.4   OLM   Chan 81   N46°58.30' W122°54.11'   at fld. 200/19E.
         VOR portion unusable:
            043°–082° blo 17,500'   333°–358° byd 25 NM blo 6,000'
            143°–183° byd 20 NM blo 6,000'   358°–043° blo 10 NM blo 6,000'
            333°–358° byd 10 NM blo 4,000'   358°–043° byd 20 NM blo 7,000'
         DME unusable:
            223°–258° byd 20 NM blo 4,100'   358°–043° blo 10 NM blo 6,000'
            258°–283° byd 30 NM blo 4,100'   358°–043° byd 20 NM blo 7,000'
      ILS 111.9   I–OLM   Rwy 17   Unmonitored during hours twr closed. BC unusable.
      COMM/NAVAID REMARKS: Emerg frequency 121.5 not avbl at twr.

Figure 4-4. Tacoma Narrows, Olympia, WA—excerpts from A/FD

Figure 4-5. Bowers Field, Ellensburg, Washington

A magenta halo around an airport symbol has special meaning to VFR pilots. That magenta circle means that the floor of controlled airspace (Class E) dips to 700 feet above the ground and, not coincidentally, that the airport has at least one instrument approach procedure. Note the subtle trap—go up from the surface 699 feet and you remain in Class G airspace with its daytime rule of "clear of clouds." You can fool around (and I use the phrase advisedly) beneath the clouds doing touch-and-goes with a mere mile of flight visibility. Gain one more foot of altitude and you are in Class E airspace, honor bound to stay 500 feet below, 1,000 feet above, and 2,000 feet laterally from all clouds with three miles visibility. Why? Because a pilot flying on instruments can pop out of one of those clouds and surprise you. Ellensburg, airport E on the sectional chart (and Figure 4-5), is an example.

The first rule of operations for VFR pilots at such an airport is to stay on the ground unless the base of the clouds is at least 1,500 feet above the airport. (You have to be 500 feet below the clouds for pattern work, right?) That will give an approaching instrument pilot a clear view of the airport long before the airplane leaves controlled airspace, letting you do your takeoff and landing practice in safety.

The second rule is a little more complicated. If the airport has a UNICOM or no radio at all, you want to be on the appropriate frequency to make your position reports in the pattern and to hear other pilots in the area. At the same time, it would be helpful if you could listen to the appropriate air traffic control frequency to hear any transmissions from inbound pilots who are shooting the instrument approach. Weather has nothing to do with this, by the way—pilots working on their instrument ratings frequently practice approaches in VFR weather while wearing a hood or other view-limiting device.

This is the way it should work: The ATC controller tells the instrument pilot (on the ATC frequency),

> **CONTROLLER** *"Cleared for the approach, change to advisory frequency approved."*

That tells the instrument pilot to switch to UNICOM or MULTICOM to tell pilots like you that they are coming your way and most likely will make a straight-in approach instead of flying a rectangular pattern. Because you have one receiver tuned to the ATC frequency you are ahead of the game—you know what to expect.

An instrument pilot who is not too sharp will switch over to UNICOM or MULTICOM and say something like,

> **PILOT** *"Podunk traffic, Baron 1014W at RIBIT, low approach."*

(RIBIT is an example of a fix or intersection in an instrument approach procedure.) You, of course, being neither an instrument pilot nor in training for the rating, have no idea what RIBIT is. It gives you no clue at all as to what is going on, and probably makes you mad if your instructor has drummed into your head the incorrect information that straight-in approaches are a no-no. Tell your instructor to read Advisory Circular 90-66A (and look back at Figure 4-2).

An instrument pilot or student who is more on the ball might say:

> **PILOT** *"Podunk traffic, Baron 70497 five miles out, straight in on an instrument approach to runway 19."*

If you regularly fly into airports with instrument approaches it might be a good idea to get together with your instructor (who is instrument-rated, of course) and talk about how the instrument folks operate in the vicinity of any nontowered airports you use frequently.

Summary

As a practical matter, you are not going to spend much time in Class G airspace — it is too close to the ground. The only (dubious) advantage of Class G airspace is the relaxed daytime VFR visibility and cloud clearance requirements. Flying with only one mile visibility and clear of clouds is rightly dubbed "scud running" and doing so has led to many accidents attributed to "inadvertent VFR flight into instrument meteorological conditions." At night, the requirements for VFR flight in Class G airspace are much more stringent.

Still, the communication situations we have discussed are valid and will be found to apply equally well in Class E airspace.

One final anecdote about the use of radios at nontowered airports:

Two retired pilots (airline or military or both, I can't remember which) had worked separately for hundreds of hours each restoring antique biplanes. One Sunday morning each of them decided to go flying from a nontowered airport in Class G airspace. The windsock was hanging straight down. One taxied to the east end of the runway, the other to the west end — these airplanes are notorious for poor pilot visibility and neither saw the other airplane taxiing. Because of a hump in the center of the runway, neither pilot could see the other as they moved into position for takeoff. When they saw that they were going to collide in the middle of the runway, each veered to the right and there was nothing but torn fabric and broken ribs where the left wings used to be.

Both airplanes had radios, but neither pilot had said a word because "at an uncontrolled field you don't have to use your radio." I'm sure both pilots reconsidered that logic as they totaled up the damage.

It's Your Typical Flight Environment

A flight from one small airport to another, at the relatively low altitudes of Class G airspace, places few communications demands on you. Checking in with the air traffic control system is usually a waste of time because their radar coverage is minimal to nonexistent close to the ground. Tree-top flying in light airplanes (and I mean the really light ones, like ultralights and light-sport aircraft) is fun and won't result in a lot of noise complaints if the flights take place away from population centers. However, serious cross-country flying means getting into Class E airspace.

No matter what class of airspace you are departing from or enroute to, there is no requirement that you communicate with anyone, and there are some hardy souls who use that excuse for ignoring their radios for most of the flight. That is not smart. In Chapter 10 you will learn about flight following and pilot reports; in this chapter you will learn about Flight Watch. All are enroute activities.

Because there are so many combinations of departure and arrival airspace classes, you will find yourself jumping around in this book. For example, if you are departing an uncontrolled airport and going to an airport in Class B airspace, read Chapters 5 and 6 for departure and Chapter 8 for arrival.

Because Class E airspace exists from 700 feet AGL or 1,200 feet all the way to 18,000 feet MSL, there is a tremendous volume of it in the United States and you will spend most of your flying hours buzzing around in it. When you crossed the magenta tint line between Sky Harbor and Harvey Field, the floor of Class E airspace dropped from 1,200 feet to 700 feet above the ground. Unless you were flying low enough to have people report you to the FAA for unwarranted low flying, you entered Class E airspace at that moment.

We've flown from Sky Harbor to Harvey Field and have used plane-to-plane and UNICOM frequencies. Your friend Joe is going back to Sky Harbor. It's time to stretch your wings and fly cross-country from Harvey Field to Olympia (Figure 5-1, and airport C on the sectional chart), about 65 miles away. The smart thing for you to do now is to go into the pilot lounge to prepare a flight plan and file it by telephone. Even if Harvey Field had a Remote Communications Outlet, filing on the radio when a phone is available is poor practice. Dialing 1-800-WX-BRIEF will do the trick no matter where you are.

It's too early in the book to get involved with Class B or D airspace, so your flight to Olympia will be via airways: direct to the Paine Field VOR, then V-287 to OLM. With the exercise of a little imagination and application of the regulations, you will be able to remain in Class E airspace all the way to the Olympia area.

Departing Harvey Field

Harvey Field has a paved runway (and a grass runway) with a taxiway leading from the gas island to the runup area. You could fire up your radio after starting the engine and call the UNICOM operator to learn which runway is in use, but haven't you had an opportunity to check the windsock and to monitor landing and departing traffic while preflighting your airplane? If that hasn't been the case (your airplane might be in between two rows of hangars) you could taxi out to a position with a direct line-of-sight to the FBO office and say:

> PILOT *"Harvey UNICOM, Piper 70497, request runway in use."*

At locations with parachuting activity, such as Harvey, it might be a good idea to ask the UNICOM operator about any planned jump activity; the situation may have changed since you arrived.

With the runway information in hand, and considering that airplanes might be taxiing out from other tie-down areas or between rows of hangars, it is a good idea to warn others of your intentions by transmitting:

> PILOT *"Harvey traffic, Piper 70497 taxiing from the ramp to runway 32."*

Many small airports such as Harvey Field have a tetrahedron in addition to one or more windsocks, and it is not unusual for airport management to tie the tetrahedron in position to indicate the preferred runway to a commuter or regional airline flight. Never base your plans on the position of a tetrahedron — go by the windsock (*see* Figure 5-1).

Figure 5-1. Runway indicators

Once you have warned any potentially conflicting ground traffic of your intended taxi route, just keep your eyes open and your finger off of the push-to-talk switch until you have completed your runup and have checked the skies visually for landing traffic. Then,

PILOT *"Harvey traffic, Piper 70497 departing runway 32, [standard, downwind, straight-out] departure"*

—will do the trick. When you are airborne, don't be in a hurry to switch away from the UNICOM frequency; if you are westbound, and an incoming pilot calls Harvey traffic to report five miles west, that is information you want to hear.

Now it is time to open that flight plan you filed on the telephone with the Seattle Flight Service Station:

> **PILOT** *"Seattle Radio, Piper 70497, 122.55 over."*

(See the FSS frequency above the PAE VOR frequency box?)

> **RADIO** *"497, Seattle Radio."*

> **PILOT** *"Seattle Radio, Piper 70497, please open my VFR flight plan from Harvey Field to Olympia via Victor 287 at ten minutes past the hour."*

> **RADIO** *"497, Seattle Radio, your flight plan is open. Seattle altimeter 30.12. Have a nice flight."*

That exchange is perfectly proper, but you could spend less time on the air by doing it all in one transmission:

> **PILOT** *"Seattle Radio, Piper 70497 on 122.55, please open my VFR flight plan from Harvey Field to Olympia at ten minutes past the hour."*

The AIM says that the initial callup can be omitted if radio reception is reasonably assured, and at a large automated flight service station (AFSS) such as Seattle's, with lots of specialists, this would probably work. At a more remote FSS (the kind that are being shut down), the call-up and response method might work better because chances are that the lone specialist is on the phone or otherwise busy.

You have now laid the groundwork for a search effort if you should fail to arrive at Olympia fairly close to your estimated time of arrival. The search-and-rescue folks don't like to be forced to spread their efforts over thousands of square miles of terrain if you do end up missing, and you can help them (and yourself) quite a bit by making position reports. For some reason this is becoming a lost art among pilots, and FSS specialists almost seem surprised when they receive one; but it is a wise practice for you to follow. Looking at the sectional, it would appear that the Apex airport (near Poulsbo on V287) would be a good reporting point. But let's get on our way to Olympia.

Over the Top of Paine Field

You opened your flight plan as soon as you could safely do so after leaving the immediate vicinity of Harvey Field; now, because you have filed a route direct to the Paine Field VOR (Figure 5-2, and airport F on the sectional), you have a decision to make. When Paine's control tower is in operation (see the asterisk indicating

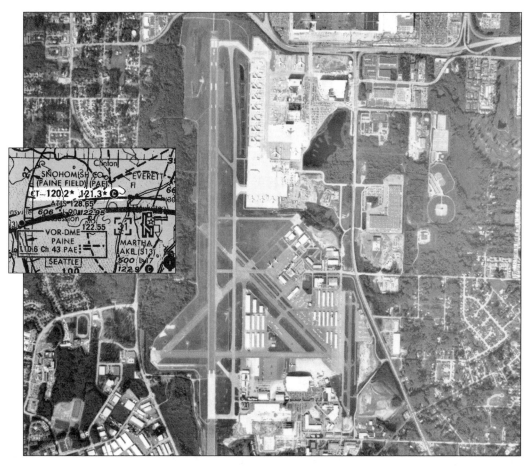

Figure 5-2. Paine Field, Everett, Washington

that it is a part-time tower?), Class D airspace is in effect inside that blue dashed line, and Class D airspace doesn't show up until the next chapter!

Just to stay on the subject of this chapter, I suggest that you fly over the Class D airspace to remain in Class E airspace. Paine's field elevation is 606 feet above sea level, and Class D airspace extends 2,500 feet above ground level, so you need to be at least 3,107 feet MSL to clear it. It is only three nautical miles from Harvey Field to the boundary of Paine's Class D airspace, and I doubt that most small planes based at an airport like Sky Harbor can climb 1,000 feet per minute. So just for the purpose of the exercise you will circle up to 3,000 feet over the Snohomish

Valley west of Harvey Field, before heading west and climbing to 4,500 feet, the proper altitude for your direction of flight (you don't have to observe the hemispherical rule while climbing or descending).

What if the base of the clouds was at 2,500 feet? What if you just didn't want to spiral up to 3,000 feet? Would you go around the Class D airspace? I sure hope not. It only takes a moment to say on 120.2 MHz (Paine's control tower frequency):

> **PILOT** *"Paine Tower, Piper 70497, I'd like to cross the field east to west at 2,500 feet."*

> **TOWER** *"Piper 70497, Paine Tower, cleared as requested, report clear."*

When you are over Paine Field you can tune your VOR receiver to 110.6 MHz and put the Omnibearing Selector on 236° for course guidance along V-287.

Now that didn't hurt a bit, did it? As soon as you reach the west side of the dashed blue line, simply say,

> **PILOT** *"Paine Tower, 497 is clear to the west."*

—and go on your way. When the tower is closed, you don't have to talk to anyone because the Class D airspace becomes Class G from the surface up to 700 feet AGL—but you knew that because you looked it up in the Airport/Facility Directory.

Extensions

Notice that Paine Field (and Olympia) have magenta dashed lines abutting the Class D airspace. These indicate Class E extensions, designated to protect instrument flights inbound to the airport. As is the case in other Class E airspace, there is no communication requirement for flight through the extensions. However, regulations or no regulations, it doesn't make good sense to fly silently through an area which you know is used by inbound instrument pilots. If you decide to fly around Class D airspace to avoid talking to the tower, do not fly through a Class E extension. It's just common sense.

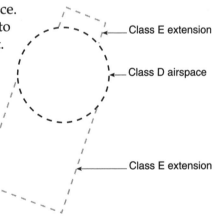

Figure 5-3. Class E extension

Making Position Reports

As your cross-country courseline takes you close to remote communication outlets or airports with flight service stations on the field, just give them a call on the frequency shown on the chart (remember to include the frequency in your initial call) and, after they have acknowledged your call, say:

> **PILOT** *"Seattle Radio, Piper 70497 on 122.5 [or 123.65*] over Apex at 45 past the hour on a VFR flight plan to Olympia."*

* see the RCO frequency box just southeast of Apex?

The FSS specialist will acknowledge your call and give you the current altimeter setting. At the same time, the specialist will log your call and the time of reception, and if you fail to land at the appointed time, the authorities will know that you got at least as far as Apex.

EXCERPT FROM CALLBACK NEWSLETTER

"WITH A CEILING OF 700 FEET, A GA AIRCRAFT WAS PRACTICING PATTERN WORK AT WHAT HE SAID WAS 500 FEET AGL. I CONSIDER IT UNSAFE WHEN WE BREAK OUT RIGHT AT MINIMUMS ON THE CIRCLING APPROACH, SWITCH FREQUENCIES FROM APPROACH TO UNICOM AND HEAR SOMEONE SAY HE WILL DO A 360 ON CROSSWIND TO AVOID THE JET. WE COULD ALL HAVE BEEN DEAD RIGHT."

Radar Flight Following

Do you wear both a belt and suspenders? If that is the case, or if you just like to take advantage of the services your tax dollars have purchased, by all means use radar flight following in addition to filing a VFR flight plan. When your flight is being followed by radar, you can forget about making position reports. ATC will advise you of traffic, but if you want vectors to avoid a problem you must ask for them. Flight following is a tool that all VFR pilots should take advantage of.

Unless you are departing from an airport that has a terminal radar facility — and that would most likely be one in Class B, C, or D airspace — you will make your request for radar flight following to an air route traffic control center, called Center for short. You will find Center frequencies in the back of the Airport/Facility Directory or by calling an FSS:

PILOT *"Seattle Radio, Piper 70497 over Paine Field—what is the Center frequency in this area?"*

RADIO *"Piper 70497, Seattle Radio, contact Seattle Center on 125.1 in that area."*

PILOT *"125.1, thank you, 497."*

PILOT *"Seattle Center, Piper 70497 over Paine Field at 2,500 feet, request flight following."*

CENTER *"497, Seattle Center, squawk 3456 and ident [pause] radar contact, advise any altitude change."*

The controller might transmit "Say type aircraft," because you did not specify the type in your initial contact. There are many different models of general aviation aircraft, and the controller expects the correct type identifier from Appendix A to the *Air Traffic Control Handbook* (FAA Order 7110.65).

You must understand that all controlled airspace is "owned" by the various Air Traffic Control Centers, who release little doughnuts or wedding cakes or whatever shapes of airspace to Approach Controls and towers. Both Center and Approach facilities subdivide their airspace into sectors, each with its own frequency and controller... and you have no way of knowing where the sector boundaries are. Further, they sign agreements to share responsibilities, so you might find yourself talking to a Center controller at night over a location where you talked to Approach in the daytime. Or in low-activity periods, sectors might be combined, so you talk to one controller today where you talked to two controllers yesterday. Don't let it bother you...it's just the way the system works.

When departing from a tower-controlled airport, tell the local controller when you report ready for takeoff:

PILOT *"Galveston Tower, Baron 1014W is ready for takeoff, VFR San Antonio, 6,500 feet, request handoff for radar flight following."*

The local controller will check with the radar folks and, workload permitting, you will get a discrete transponder squawk that will almost certainly begin with a zero, if it comes from a terminal radar facility (approach control). You will hear this after you have taken off:

TOWER *"Baron 1014W, contact Departure 134.45."*

Must you be departing a major airport with a radar facility to ask for flight following? Look at Figures 5-5 and 6-2. These are not large airports, and neither has a radar facility. The A/FD for both, however, shows an Approach/Departure frequency. Take a look at the listing for airports you use and see if they show availability of a radar facility frequency. Unfortunately, when you near the edge of the terminal radar facility's airspace, you will hear,

CONTROLLER *"Radar service terminated, squawk VFR. Good day."*

If approach control has requested a squawk code from the air route traffic control computer, however, the code will *not* start with a zero and you will keep the same code throughout your flight. When you approach the boundary of terminal radar airspace you will be handed off to a Center controller.

Handoffs

Controllers exchange information by landline, so a lot goes on behind the scenes that you are unaware of. When you approach the boundary of a controller's airspace, he or she causes your identification block on the receiving controller's scope to blink. The receiving controller takes action to accept your flight, and the block stops blinking. Then and only then does the controller you are talking to say "Contact Atlanta Center on 134.5." The new controller knows who you are, where you are going, your aircraft type, and your altitude. They don't believe in altitude reporting transponders, though, so when you contact the receiving controller you should say:

PILOT *"Atlanta Center, Baron 1014W, level seven thousand."*

But what if the receiving controller is overloaded and cannot accept a VFR flight and radar service is terminated? Squawk VFR as directed and continue just like any other VFR flight...but after a few miles put in your own request:

PILOT *"Atlanta Center, Baron 1014W, a BE-55, 20 miles northeast of Sugarloaf Mountain, level seven thousand five hundred, request VFR flight following to Richmond, Virginia."*

Don't tell the controller that you have been handed off from a previous controller, because you haven't...and the controller you are talking to has no clue that you are coming.

You should hear,

CENTER *"Baron 1014W, radar contact two zero miles northeast of Sugarloaf Mountain, maintain VFR and advise prior to any altitude changes."*

I suggest that you either buy an IFR low altitude en route chart that covers your area of operations, or ask an instrument-rated friend to give you an expired chart instead of recycling it. Learning to find Center frequencies on low altitude charts will take about two minutes, max.

You must understand that the sole duty of air traffic controllers is providing separation between IFR flights. Any service provided to a VFR flight is on a workload-permitting basis, and if the controller's response to your request for flight following is "Unable," you will just have to do without. The controller might be swamped with work or just having a bad hair day—it doesn't make any difference.

Every terminal radar facility (near big airports) is "Approach Control," so use that as a callsign. When you are receiving radar flight following service, don't worry about getting clearance to fly through the Class D airspace of a tower-controlled airport along your route—it is the radar controller's responsibility to coordinate your flight with the tower.

Once you are in the system, each Center controller will arrange for the controller in the next sector to pick you up and will then tell you to contact that controller on a new frequency. It will still be Center, but a different frequency and usually (but not always, because of remote radar sites), a different voice. Occasionally, Center will lose radar contact because of your distance or altitude and will say:

> **CENTER** *"Baron 1014W, radar contact lost, squawk VFR, change to advisory frequency approved."*

My recommendation in that situation is this:

> **PILOT** *"Fort Worth Center, Baron 1014W would like to stay on this code and contact the next sector when in range."*

That *might* bring,

> **CENTER** *"Baron 1014W continue present squawk, contact Center on 134.8 in ten miles."*

Of course, it might result in "unable," too.

Note that there is a difference between "radar service terminated" and "radar service lost." Service is *terminated* at the controller's option. Radar contact is *lost* because of altitude or obstacles...you will get the service back as soon as contact is regained. Do not squawk VFR and change frequencies when radar contact is lost...or any time, without instructions to do so.

East of the Mississippi, the heavy traffic and complex airway system makes flight following a sometime thing. Your chances of receiving this service are greater in the west, but it still breaks down to controller workload, and whether or not the controller feels that providing radar service to you will adversely impact his or her primary duty to the folks on IFR flight plans.

Terminating Radar Service

Never leave an ATC frequency without informing the controller. You might think that the controller is too busy, or that you would just be a bother, but consider this…if you simply disappear from the frequency the controller *must* assume the worst and initiate search-and-rescue procedures. By saying,

PILOT *"Denver Center, Baron 1014W terminating flight following, thanks for your help."*

CENTER *"Roger, 14W, squawk 1200 (or squawk VFR), frequency change approved."*

The controller assumes that you are going to switch to UNICOM, Flight Watch, etc.

Requests and Clearances

As you cruise along using radar flight following in Class E airspace (which will be most of the time), you are the master of your destiny. If you want to change altitude you are free to do so…but just in case there is a B-2 bomber at an intermediate altitude talking to the radar controller on a military frequency, you should advise ATC before making the change:

PILOT *"Oakland Center, Cessna 1357X leaving 4,500 for 6,500."*

You do not "Request altitude change," because you are not at an assigned altitude and the controller has no authority to assign an altitude to a VFR flight.

Similarly, you need not "Request clearance to…" unless you are approaching Class B airspace—Classes C and D require only that communications be established, not that a clearance be issued. Just tell the controller what you want to do—it is not necessary to ask for permission.

Strange Field Entry

As you approach any airport you have a responsibility to other pilots in the area to apprise them of your position and your intentions. Listen on the Common Traffic Advisory Frequency (CTAF) for reports from other pilots already in the pattern; you can't go wrong by going along with the flow. In any event, when you are about ten miles from the airport, transmit:

PILOT *"Podunk area traffic, Piper 70497 is ten miles northeast for touch-and-goes at Podunk."*

Do not say "Any traffic in the area please advise." The AIM specifically cautions against the use of this phrase.

Then make position reports on the 45, downwind, base, and final if you think they are appropriate.

This accomplishes two things: It lets others know that you are going to join the pattern, and it gives them an indication of where to look for you. If you hear someone say something on the frequency that implies a possible conflict, speak up!

> **PILOT** *"Piper 497 is on the 45."*

> **PILOT** *"Piper 497, Baron 1014W turning crosswind to downwind does not have you in sight."*

Or,

> **PILOT** *"Podunk traffic, Baron 1014W is straight in for runway 19, practice NDB approach."*

> **PILOT** *"Baron 1014W, Piper 40497 is on base for 19, has you in sight; we'll be touch-and-go ahead of you."*

Departing a Strange Field in Class E Airspace

There is no substitute for advice from the locals. If you are departing, ask one of the instructors (or a fueler, or the person at the counter) if there are any special departure procedures you should be aware of. As an arrival, if you have researched the A/FD there are probably no surprises. However, it is not unusual for an itinerant pilot to scan the airport bulletin board and learn that the local airport authority has an ordinance that he or she violated on the way in.

No Radio (NORDO)

Although it may be hard to believe in this enlightened age, there are airplanes flying around in the friendly skies without radios. Another form of NORDO is the airplane

Color & Type of Signal	Movement of Vehicles, Equipment & Personnel	Aircraft on the Ground	Aircraft in Flight
Steady green	Cleared to cross, proceed or go	Cleared for takeoff	Cleared to land
Flashing green	Not applicable	Cleared for taxi	Return for landing (to be followed by steady green at the proper time)
Steady red	STOP	STOP	Give way to other aircraft and continue circling
Flashing red	Clear the taxiway/runway	Taxi clear of the runway in use	Airport unsafe, do not land
Flashing white	Return to starting point on airport	Return to starting point on airport	Not applicable
Alternating red and green	Exercise extreme caution	Exercise extreme caution	Exercise extreme caution

Figure 5-4. Light signals for NORDO situations.

with an instructor who has turned the radio volume down because it was interfering with his or her comments to the student. The assumption that a quiet frequency means there is no traffic in the pattern can be quite dangerous. (*See* Figure 5-4.)

Special Visual Flight Rules

When an airport in Class E airspace has an instrument approach and also has full-time weather observers, Class E airspace begins at the surface; the boundaries of its surface area are shown by dashed lines. Bremerton National Airport (Figure 5-5, and airport G on the sectional chart) is a good example. Bremerton National does not have a control tower, as indicated by the magenta color of the airport

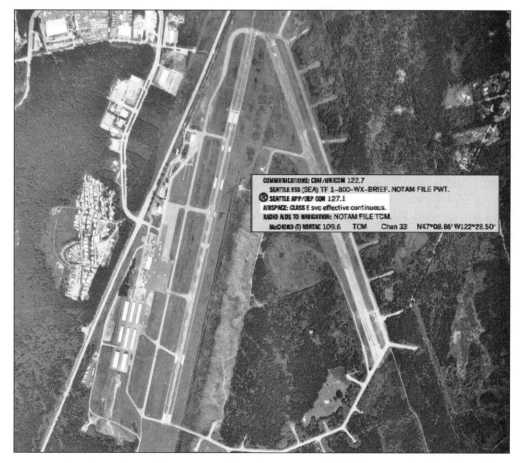

Figure 5-5. Bremerton National Airport, Washington

symbol, but the Airport/Facility Directory indicates that Class E airspace is in effect continuously.

Note that a digital A/FD ("d-A/FD") is available online at http://www.naco.faa. gov/index.asp?xml=naco/online/d_afd for use during preflight planning.

A non-instrument-rated pilot (or an instrument-rated pilot who for some reason does not want to file an instrument flight plan), who wants to take off from or land at Bremerton National when the ceiling is less than 1,000 feet or the visibility is less than 3 miles, must receive a special visual flight rules (SVFR) clearance from ATC. Nothing on the chart indicates that a pilot who wants a SVFR clearance to depart Bremerton has any option other than to file on the telephone by calling 1-800-WX-BRIEF.

A pilot who intends to land at Bremerton and learns that the field is below basic VFR minimums (by listening to the automated weather observation system (AWOS) on 121.2) can either call the Seattle FSS on 122.5/123.65 and request a SVFR clearance, or call the Seattle Approach/Departure Control radar facility directly on 127.1 MHz. Where did I get that frequency? It sure isn't on the sectional. I know that you will think that I get a royalty from the National Aeronautical Charting Office for each copy of the Airport/Facility Directory they sell, but that's the only place a VFR pilot will find that information.

A Special VFR clearance is something new to the average VFR pilot. It contains a transponder code to squawk, and precise directions on how to enter or leave the Class E surface area; and it must be read back to the controller verbatim. Before you ask for a SVFR clearance, have a piece of paper and something to write with at hand.

CONTROLLER *"ATC clears Piper 70497 out of the Bremerton National Airport Echo surface area to the east" [note the mountains and the antenna to the west], "maintain Special VFR conditions at or below 1,500 feet until clear. Report clear of the Echo surface area."*

Or,

CONTROLLER *"ATC clears Piper 70497 into the Bremerton Echo airspace from the north at or below 1,500 feet. Report landing on 127.1."*

It is important for you to realize that a Special VFR clearance to depart Class E airspace (or C/D when appropriate) just gets you across the blue dashed line depicting the airspace boundary. Once past that line, you are on your own...and the weather had better be good VFR. Imagine how infrequently there will be a low ceiling and/or restricted visibility around an airport and clear skies (or, at least, better weather) five miles away.

Beneath the Ceiling

A Special VFR clearance is required in order to operate *beneath the ceiling* while in Class E airspace when the weather is below VFR minimums. Pilots flying over a tower-controlled airport far above the tower's airspace sometimes wonder if they have to call the tower—the answer is "no."

Summary

Unless you like to fly at relatively low altitudes, you will do most of your serious cross-country flying in Class E airspace—hopscotching between the other classes of airspace. Pay attention to the visibility and cloud clearance requirements, and set minimums for yourself that are higher than those required by the regs.

When is a Tower Not a Tower?

An airport in Class D airspace is shown on a sectional chart with a blue airport symbol; on the sectional chart excerpt there are several: Paine Field, SeaTac, Boeing Field, Renton, Olympia, etc. The Class D airspace extends from the surface to 2,500 feet above airport elevation (in most cases), and its surface area is shown with a blue dashed line. Where there is no dashed line, you will see a magenta-tint circle indicating that the floor of controlled airspace is at 700 feet AGL. Class D airspace exists *only* when the tower is in operation, however.

Note that I said Class D airspace begins at the surface, not at field elevation. Look at Paine Field, airport F on the sectional chart. The little "31" in a square indicates that the top of Paine's Class D airspace is 3,100 feet above sea level. But what if you are flying 500 feet above the water just west of Paine—aren't you beneath the Class D airspace because you are lower than field elevation? Absolutely not—Class D airspace starts at the surface, down where all those salmon are, and you must be in communication with the tower even if you are flying below field elevation.

Some towers release the top 500/1,000 feet of their Class D airspace to the ARTCC or Radar Approach Control facility whose airspace overlies it. In most cases, if you request permission from the tower controller to overfly the airport at those altitudes, the controller will coordinate with the radar facility and you won't have to say a word. In very few situations (usually when the radar facility has an airplane on an instrument flight plan in the vicinity) you will be given the frequency for that facility and told who to contact.

It is not unusual to hear "cross at midfield" when you approach the airport from the side opposite the traffic pattern. In the absence of a specific altitude assignment from the controller, fly across 500 feet above pattern altitude. Get traffic advice from the controller and keep your eyes peeled.

The local controller's authority is limited to that cylinder of airspace indicated with blue dashed lines. You must establish two-way radio communication with that controller for authorization to operate within that cylinder. That means, of course, that you must call for authorization well before you reach the dashed line on the chart that marks the boundary. If you are unable to make contact due to a busy frequency, you must circle around outside of the Class D airspace until you do make radio contact. (NORDO operations will be discussed later.)

It is important to note that tower controllers are not required to and do not have the authority to provide separation services between VFR aircraft. You will hear what appear to be control-type transmissions ("Follow the Cessna ahead on downwind"; "You are number two to land after the Beech twin on base, report that aircraft in sight"), but actual separation is the responsibility of VFR pilots.

The shape of the surface area of Class D airspace varies widely, but is always as depicted on sectional charts. To protect inbound IFR flights, the airspace planners create extensions along the inbound courses of instrument approach procedures. Where such an extension is two miles in length or shorter, it is still Class D airspace and you must talk to the tower controller to fly through it. Where an extension is more than two miles in length the excess becomes Class E airspace and you can fly through it without talking to the tower—but why would you want to? For that matter, why would you want to fly across the final approach course of an instrument procedure unless you knew for sure that no one was using it?

Class D airspace reverts to Class E or G when the last controller pulls the plug on the coffee pot and goes home. So you have to know the tower's operating hours before you pick up the mike to call for entry into Class D airspace, because you might not need permission. Sectional charts have an asterisk next to the tower frequency when it operates only part-time. The Airport/Facility Directory is a better source of information than any chart, because it gives details on hours of operation and the status of the airspace when the tower is closed (*see* Figure 6-1).

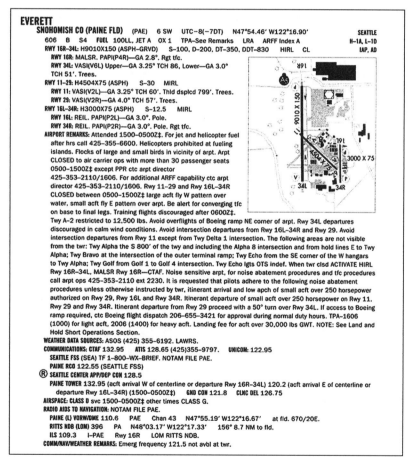

EVERETT
SNOHOMISH CO (PAINE FLD) (PAE) 6 SW UTC–8(–7DT) N47°54.46' W122°16.90' SEATTLE
 606 B S4 FUEL 100LL, JET A OX 1 TPA–See Remarks LRA ARFF Index A H–1A, L–1D
RWY 16R–34L: H9010X150 (ASPH–GRVD) S–100, D–200, DT–350, DDT–830 HIRL CL IAP, AD
 RWY 16R: MALSR. PAPI(P4R)—GA 2.8°. Rgt tfc.
 RWY 34L: VASI(V6L) Upper—GA 3.25° TCH 86, Lower—GA 3.0° TCH 51'. Trees.
RWY 11–29: H4504X75 (ASPH) S–30 MIRL
 RWY 11: VASI(V2L)—GA 3.25° TCH 60'. Thld dsplcd 799'. Trees.
 RWY 29: VASI(V2R)—GA 4.0° TCH 57'. Trees.
RWY 16L–34R: H3000X75 (ASPH) S–12.5 MIRL
 RWY 16L: REIL. PAPI(P2L)—GA 3.0°. Pole.
 RWY 34R: REIL. PAPI(P2R)—GA 3.0°. Pole. Rgt tfc.
AIRPORT REMARKS: Attended 1500–0500Z‡. For jet and helicopter fuel after hrs call 425–355–6600. Helicopters prohibited at fueling islands. Flocks of large and small birds in vicinity of arpt. Arpt CLOSED to air carrier ops with more than 30 passenger seats 0500–1500Z‡ except PPR ctc arpt director 425–353–2110/1606. For additional ARFF capability ctc arpt director 425–353–2110/1606. Rwy 11–29 and Rwy 16L–34R CLOSED between 0500–1500Z‡ large acft fly W pattern over water, small acft fly E pattern over arpt. Be alert for converging tfc on base to final legs. Training flights discouraged after 0600Z‡. Twy A–2 restricted to 12,500 lbs. Avoid overflights of Boeing ramp NE corner of arpt. Rwy 34L departures discouraged in calm wind conditions. Avoid intersection departures from Rwy 16L–34R and Rwy 29. Avoid intersection departures from Rwy 11 except from Twy Delta 1 intersection. The following areas are not visible from the twr: Twy Alpha the S 800' of the twy and including the Alpha 8 intersection and from hold lines E to Twy Alpha; Twy Bravo at the intersection of the outer terminal ramp; Twy Echo from the SE corner of the W hangars to Twy Alpha; Twy Golf from Golf 1 to Golf 4 intersection. Twy Echo lgts OTS indef. When twr clsd ACTIVATE HIRL Rwy 16R–34L, MALSR Rwy 16R—CTAF. Noise sensitive arpt, for noise abatement procedures and tfc procedures call arpt ops 425–353–2110 ext 2230. It is requested that pilots adhere to the following noise abatement procedures unless otherwise instructed by twr, itinerant arrival and low apch of small acft over 250 horsepower authorized on Rwy 29, Rwy 16L and Rwy 34R. Itinerant departure of small acft over 250 horsepower on Rwy 11, Rwy 29 and Rwy 34R. Itinerant departure from Rwy 29 proceed with a 50° turn over Rwy 34L. If access to Boeing ramp required, ctc Boeing flight dispatch 206–655–3421 for approval during normal duty hours. TPA–1606 (1000) for light acft, 2006 (1400) for heavy acft. Landing fee for acft over 30,000 lbs GWT. NOTE: See Land and Hold Short Operations Section.
WEATHER DATA SOURCES: ASOS (425) 355–6192. LAWRS.
COMMUNICATIONS: CTAF 132.95 ATIS 128.65 (425)355–9797. UNICOM: 122.95
 SEATTLE FSS (SEA) TF 1–800–WX–BRIEF. NOTAM FILE PAE.
 PAINE RCO 122.55 (SEATTLE FSS)
Ⓡ SEATTLE CENTER APP/DEP CON 128.5
 PAINE TOWER 132.95 (acft arrival W of centerline or departure Rwy 16R–34L) 120.2 (acft arrival E of centerline or departure Rwy 16L–34R) (1500–0500Z‡) GND CON 121.8 CLNC DEL 126.75
AIRSPACE: CLASS D svc 1500–0500Z‡ other times CLASS G.
RADIO AIDS TO NAVIGATION: NOTAM FILE PAE.
 PAINE (L) VORW/DME 110.6 PAE Chan 43 N47°55.19' W122°16.67' at fld. 670/20E.
 RITTS NDB (LOM) 396 PA N48°03.17' W122°17.33' 156° 8.7 NM to fld.
 ILS 109.3 I–PAE Rwy 16R LOM RITTS NDB.
COMM/NAV/WEATHER REMARKS: Emerg frequency 121.5 not avbl at twr.

Figure 6-1. Paine Field excerpt from A/FD

More often than not, Class D airspace becomes Class E all the way to the ground when the tower shuts down, so the visibility and cloud clearance requirements for basic VFR remain in effect. You will have to get a Special VFR clearance from ATC to enter or depart the surface area of the airspace if the ceiling is less than 1,000 feet, or the visibility is less than 3 miles. Notice that the Common Traffic Advisory Frequency is still the tower frequency, even when the tower is closed.

At those airports where the tower controllers are also the weather observers, they take the bottom 700 feet of controlled airspace home with them and leave Class G airspace behind. That means "one mile visibility and clear of clouds" in the daytime and 3 miles visibility with basic VFR cloud clearance at night (except for folks staying within one-half mile of the runway while doing pattern work). Because the airspace is uncontrolled, no SVFR clearance is required if your altitude is less than 700 feet above the ground when within the surface area defined by the blue dashed line.

Two airport examples that illustrate this point are Olympia and Tacoma Narrows, Washington (airports C and D on the sectional chart excerpt). Olympia shows a part-time tower but a full-time NWS weather observer. A Special VFR clearance from ATC must be obtained before any operation to, from, or through the Olympia surface area depicted on the sectional (beneath a ceiling of 1,000 feet or less). Tacoma, on the other hand, has no weather capability when the tower is not in operation and during those hours, controlled airspace begins 700 feet above the airport. (I warned you in the Introduction that I would repeat myself when I thought it was important to do so.) You must refer to the Airport/Facility Directory for this information (*see* the example in Figure 6-2).

Somewhere near every blue airport symbol on the sectional you will find either "See NOTAMs/Directory for Class D eff hrs"…"See NOTAMs/Directory for Class D/E (sfc) eff hrs"…or, infrequently, "See NOTAMs/Directory for Class D/G eff hrs." And just which Directory do you think these notes refer to? The A/FD, of course. If you ever have a question about what frequency to use or who to talk to, you will find the answer in the A/FD. If the information on the sectional chart and the airport listing in the A/FD do not agree, the Directory rules. Sectionals are issued every six months, while A/FDs come out every 57 days — guess which one is more likely to be current? And read the Directory's "Legend" pages, too; they contain valuable information not available elsewhere.

Figure 6-2. Tacoma Narrows Airport, Washington

```
TACOMA NARROWS   (TIW)   4 W   UTC-8(-7DT)   N47°16.08' W122°34.69'                    SEATTLE
    292   B   S4   FUEL 80, 100, 100LL, JET A   OX 4   TPA—1300(1008)   LRA          H-1A, L-1D
    RWY 17-35: H5002X150 (ASPH-AFSC)   S-50, D-80, DT-80   MIRL                          IAP
       RWY 17: MALSR. PAPI(P4L)—GA 3.0°. TCH 50'. Trees. Rgt. tfc.        RWY 35: REIL. VASI(V4L)—GA 3.0° TCH 51'.
    AIRPORT REMARKS: Attended 1600-0400Z‡. Deer on and in vicinity of arpt. Noise sensitive arpt, for noise abatement
       and tfc procedures call arpt manager 206-591-5759 or 206-851-3544. Rwy 35 REIL out of svc indefinitely.
       ACTIVATE MALSR Rwy 17 and PAPI Rwy 17—CTAF. Landing fee acft over 12,500 lbs.
    WEATHER DATA SOURCES: LAWRS
    COMMUNICATIONS: CTAF 118.5    ATIS 124.05 (1600-0400Z‡)    UNICOM 122.95
       SEATTLE FSS (SEA) TF 1-800-WX-BRIEF. NOTAM FILE TIW.
    ℝ SEATTLE APP/DEP CON 120.1
       TOWER 118.5 (1600-0400Z‡)    GND CON 121.8
    AIRSPACE: CLASS D svc effective 1600-0400Z‡ other times CLASS G.
    RADIO AIDS TO NAVIGATION: NOTAM FILE TCM.
       McCHORD (T) VORTAC 109.6    TCM    Chan 33   N47°08.86' W122°28.50'    308° 8.4 NM to fld. 285/22E.
       GRAYE NDB (MHW/LOM) 216    GR    N47°09.02' W122°36.29'    349° 7.2 NM to fld. NOTAM FILE GRF.
       ILS 109.1   I-TIW   Rwy 17    ILS unmonitored when twr clsd.
    COMM/NAVAID REMARKS: Emerg frequency 121.5 not avbl at twr.
```

Tower Frequencies

You will find the primary tower frequency in the airport's data block on a sectional. If there are multiple runways, or if the Class D airspace is divided into sectors, you will find the details on the back of the sectional on the back of the chart as well as in the A/FD.

Class G Tower?

The FAA occasionally deems it operationally necessary to establish temporary towers. This could be at a popular airshow or at a function like the Super Bowl. Naturally, these towers will not be depicted on sectional charts but their existence will be promulgated through Notices and Letters to Airmen (and the aviation press). Treat these temporary towers just as if they had Class D markings around them, because that is their legal effect. *See* Page 4-10 for notable exceptions (in "When is a Tower Not a Tower?").

"What's the ATIS?"

Most, but not all, tower-controlled airports provide an Automated Terminal Information Service (ATIS) broadcast either on a discrete frequency (shown on the sectional chart and in the A/FD) or a nearby radio navigation aid. The broadcast includes ceiling, visibility, wind, altimeter setting, runway in use, and any special information of importance to pilots.

Under normal conditions, the tower controllers prepare a new tape every time a new weather observation is received, usually every hour. However, if the weather is changing rapidly they will make tapes more frequently.

Each ATIS tape has a phonetic alphabet identifier, and you should use that identifier when making your initial contact with the ground controller when departing, and the tower controller when arriving:

> **PILOT** *"Boise Ground/Tower, Baron 1014W [give your position and intentions], with information ECHO."*

By doing so, you relieve the controller of the need to give you the current weather observation. Many pilots who have monitored the ATIS will tell the controller that they "have the numbers." ATC will not consider this transmission as assurance that you have received the latest ATIS. Just in case you listened to information ECHO but information FOXTROT is current when you call, it is best to include the phonetic identifier of the ATIS information in your initial transmission.

Listening to the ATIS is a special case of "reading the mail." It will tell you not only which runway is in use but will give you the wind, ceiling, and visibility at the time the tape recording was made, plus other useful information relating to the airport. You can use the ability of your communication receiver to defeat the squelch and get the ATIS information long before you fly into the area.

Tune to the ATIS frequency and turn the squelch knob until static fills your ears and listen for a weak voice signal. It might take a couple of times through the tape before you get everything you want, but by knowing the wind, ceiling, and visibility well in advance you have extra planning time. (Some radios may require that you pull on a knob or switch to a "test" position to defeat the squelch.)

ATIS EXAMPLE

BOEING FIELD AIRPORT INFORMATION ROMEO, 0351Z WEATHER 11,000 SCATTERED, 25,000 THIN SCATTERED, VISIBILITY 30. TEMPERATURE 16, DEW POINT 13. WIND 230 AT 5. ALTIMETER 29.84. ILS RUNWAY 13 RIGHT APPROACH, LANDING DEPARTING RUNWAYS 13 LEFT AND RIGHT. AIRCRAFT LANDING 13 LEFT ARE REMINDED TO FLY THE VASI. ALL VFR DEPARTURES SAY DIRECTION OF FLIGHT. ON INITIAL CONTACT ADVISE YOU HAVE INFORMATION ROMEO.

At tower-controlled airports that do not provide an ATIS broadcast, the tower controller will give you the ceiling, visibility, wind direction and velocity, and altimeter setting when you make your initial contact. If you have been monitoring the frequency (as you should) and hear this information transmitted to another airplane, you can save the controller a little time and trouble by including "I have the numbers" or "Cessna 1357X has the numbers" when you call. If you are departing from such an airport, the ground controller will do the honors.

Ground Control

The ground controller is usually one of a team of controllers in the tower cab, and ground control is a discrete frequency. Ground control frequencies are almost always 121.7, .8, or .9, so if a controller says "Contact ground point seven" you are expected to know that the frequency is 121.7. If you flip through the Airport/ Facility Directories for several areas, however, you will see that this is not a hard-

and-fast rule. You will also learn through experience that during slack times one person serves as both local ("tower") controller and ground controller, so you will either hear the same voice on both frequencies or hear the tower controller say "Taxi to parking."

The ground controller's responsibility is to keep you from running into someone while on a "movement area" of a tower-controlled airport. You can usually, but not always, identify movement areas by a double white line that separates them from the ramp or parking areas. Common sense plays a big role here. If you are in between two lines of T-hangars or in a common tie-down area you are on your own. Don't expect the ground controller to keep you from fender benders until you are ready to enter the taxiway—and don't expect the ground controller to give you any instructions until you are in a movement area. Having the rotating beacon and/or strobes on will help the ground controller find you when you make the initial call, but you should do your part by giving your location as accurately as possible:

PILOT *"Houston Ground Control, Cessna 1357X at the Icarus Flight School ramp, VFR to Waco with information HOTEL."*

PILOT *"Yakima Ground, Baron 1014W at south parking, staying in the pattern."*

PILOT *"Norfolk Ground, Piper 70497 at the Texaco pumps, taxi to the terminal for passenger pickup."*

When taxiing for takeoff, stay on the ground control frequency until you are really ready for takeoff. If the plane behind you notices that one of your baggage doors is open, or if the FBO wants you to return to the ramp to pick up the briefcase you left in the pilot lounge, the folks in the tower will attempt to call you on the ground frequency first. If you are the Good Samaritan who is second in line, the drill is,

PILOT *"Ground, Baron 1014W's rear baggage door is open."*

You aren't supposed to talk directly to another airplane on the ground control frequency, but believe me, the pilot of 14W will get the message.

If the ground controller includes a "hold short" instruction in your taxi clearance you are expected to read it back.

GROUND *"Cessna 1357X, runway 9 taxi, hold short of runway 22."*

PILOT *"Cessna 1357X is taxiing to runway 9, hold short of 22."*

Your taxi instructions will detail taxiways and intersections to be used en route to the takeoff runway. If you do not understand what is required of you, clarify with the ground controller without proceeding further. Runway incursions would not occur if pilots used their eyes as well as their ears.

Many airports have parallel runways, and you must understand that the taxiway between the two runways belongs to the tower (local) controller, not the ground controller, when both runways are in use. Instructions to hold short of the parallel runway or to taxi across it to the ramp will come from the tower controller:

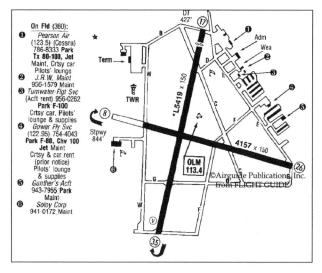

Figure 6-3. Airport diagram, Olympia, Washington

TOWER *"Baron 14W, taxi across 13R, contact ground point 9 on the far side."*

When you are going to a strange airport, nothing beats having some form of airport diagram to help you find your way around on the ground. Each volume of the Airport/Facility Directory includes airport diagrams for major airports within its coverage area, and there are a number of commercially-produced flight guides for sale at pilot supply shops. Before you walk out to your plane to depart, take a moment to study the airport diagram for your destination. Which side of the airport is the general aviation side? If you have checked the weather at the destination you know what the surface wind was at the time you called — if nothing changes, what runway will you use, and which way will you turn off to get to the general aviation ramp?

There are a few airports where the runup area is not near the hold line. At those airports, expect the ground controller to direct you to the runup area and then direct you to taxi to the hold line when you are first in line for takeoff. Listen carefully to the ATIS and ground controller instructions.

Clearance Delivery

If you see a Clearance Delivery (CD) frequency listed for your departure airport in the A/FD, understand that it is almost always intended for instrument pilots, not VFR pilots. At those airports where the tower folks want VFR pilots to contact CD because there is overlying Class C or B airspace, they will tell you so on the ATIS. After you have contacted CD and given the controller your intentions, you will be given departure instructions and will be told to contact Ground for taxi instructions.

Progressive Taxi

When at an airport you are unfamiliar with, you should ask the ground controller for "progressive taxi" instructions at your first contact after switching from the tower frequency on arrival, or on your first call when departing. After landing at such an airport you would say,

PILOT *"Knoxville Ground, Piper 70497 at intersection Bravo Two, request progressive taxi to the general aviation ramp."*

This is especially useful at night, when the airport environment seems to be a sea of blue lights. The controller will lead you by the hand:

GROUND *"Turn right there...taxi across the runway and turn left at the first taxiway on the other side..."*

The ground controller will tell you where the fixed base operations are but will not recommend one over another—that has to be your decision.

Airport diagrams are available at many websites: http://avn.faa.gov/ ap_diagrams.asp, http://www.landings.com, www.landings.com, www.naco.faa.gov, and www.aopa.org are a few.

Departures

When departing, you won't say anything to the local controller until you have completed your runup and are in position at the hold line, ready to taxi onto the active runway:

PILOT *"Boeing Tower, Baron 1014W ready for takeoff on 31 left, VFR westbound [downwind departure, staying in the pattern]."*

If you hear,

TOWER *"Baron 1014W cleared for takeoff."*

—acknowledge the clearance as you roll onto the runway.

PILOT *"Cleared for takeoff 31R, 14W."*

TOWER *"Baron 14W cleared for immediate takeoff or hold short."*

—stop and think for a moment. That clearance means that someone is on final approach, and the local controller wants you to either get moving right away or stay off the runway. The smart thing to do is to wait. Pilots who are in a hurry make mistakes. If, on the other hand, you are confident that you are really ready to go, don't pick up the mike, but push on the throttle. The delay while you reach for the mike and acknowledge the instructions might eliminate the gap in traffic that the controller was counting on. At a busy airport with ten planes in the pattern and five behind you in the runup area, be ready to go without acknowledgment but still acknowledge the clearance as soon as you can. (*See* the discussion of boom mikes in Chapter 2 on Page 2-9.)

Hold Short

Most of the time, action speaks louder than words. When the controller tells you to turn base, reaching for the microphone is a waste of time—just turn base and the controller will be able to see you complying. Grab the mike and acknowledge when the wings are level. Of all the things available for your use in the cockpit, the microphone is the least important when it comes to flying the airplane.

There is a special case that calls for a response, however. If the controller tells you to hold short of the runway your response must include the words "hold short." Clicking the mike or saying "Wilco" won't do the job.

PILOT *"Piper 497 holding short of 29"*

—this has to go on the tower's tape recorder for legal reasons.

"Position and Hold"

"Position and hold" means just that, but don't put your brain in neutral when you are sitting on the runway in takeoff position. Listen to the tower frequency—did the controller just clear someone to land on the same runway that you are parked on? If so, you should get takeoff clearance in a matter of seconds, not minutes. Actually, you won't get "taxi into position" if an inbound aircraft has been cleared to land. Wait no longer than two minutes before reminding the tower that you are holding in position.

Some pilots feel more comfortable taking a position 90 degrees to the runway so they can see the final approach course. If you want to do this, be sure to advise the controller of your intentions—the sequence of landing traffic is based on your being aligned with the runway.

You will not get a "position and hold" instruction at night; you may recall that a 737 landed on a commuter that was holding in position for takeoff at night.

Wake Turbulence

Another possible transmission from the tower controller is,

> **TOWER** *"Piper 70497 position and hold, expect three minute delay for wake turbulence."*

You can waive the delay, if you are feeling especially brave.

> **PILOT** *"497 will waive the delay."*

…will get you a takeoff clearance. After telling the local controller what you intend to do, a 90-degree turn into the wind as soon as practicable will take you away from most potential wake turbulence. Don't maintain runway heading — the turbulence descends below the generating airplane and you can't possibly outclimb a jet.

Don't anticipate what the tower controller is going to say — if you do, you will hear what you expect to hear, not what is actually said. If you are #1 at the hold line, expecting to hear,

> **TOWER** *"Position and hold,"*

and the controller says,

> **TOWER** *"Taxi across without delay and hold short on the other side,"*

…you might taxi into takeoff position on the runway through force of habit. When in doubt, ask. The 747 taxiing up behind you can wait 30 seconds while you clear things up with the controller.

Intersection Takeoffs

The tower controller will let you take off from an intersection if that is what you want to do, and in many cases you will be cleared to taxi to an intersection for takeoff. You are entirely within your rights to say,

> **PILOT** *"Baron 1014W requests the full length."*

…and in my judgment that is usually the best policy, unless you have 5,000 or 6,000 feet of runway available from the intersection. It's always nice to have runway in front of you if the engine decides to make funny noises or the door pops open just after takeoff.

When you are ready for takeoff from an intersection, be sure to include your position when calling the tower:

PILOT *"Flying Cloud tower, Cessna 1357X ready for takeoff on 9 Right at Golf."*

The tower controller has a lot of real estate to monitor and might not pick up your little airplane visually without a little help.

To make sure that there is no confusion with other aircraft making intersection departures, the controller will always include the intersection designation in transmissions.

CONTROLLER *"Baron 1014W, runway 13R at Alpha Nine, cleared for takeoff."*

Trivia note: Ground controllers never use the word "cleared," and you should not ask for a taxi clearance. The ground controller uses phrases like "taxi to runway 13," "taxi across runway 27," and so forth but does not issue clearances. Only the local controller can issue clearances.

"Request Frequency Change"

When you depart an airport with an operating control tower you are obligated to stay on the tower frequency until you cross the boundary of Class D airspace as depicted on your sectional chart. It is not necessary to request a frequency change as you leave Class D airspace, although for some reason many pilots do. This just makes additional work for the controller, who is busy enough as it is. The great majority of instructors teach their students to do this, although I can't imagine why.

If there is a possibility of conflict, the controller may say "Cessna 1357X stay with me," as you approach the boundary of Delta airspace, following up with "change to advisory frequency approved" when the conflict has been resolved.

You want to think long and hard before requesting a frequency change while still within the Class D surface area as indicated on the Sectional Chart; the tower might need to warn you of conflicting traffic. The usual terminology is,

CONTROLLER *"Change to advisory frequency approved."*

—leaving the decision up to you as to whether you want to ask ATC for flight following or shift to UNICOM. Of course, you don't have to talk to anyone while cruising along in Class E airspace.

Blocked Frequency

Very infrequently you will be faced with a situation where, because of a stuck mike on the frequency or other equipment failure, you are unable to make contact with the local or ground controller. Use your knowledge of the airport or the A/FD to get around this type of problem:

PILOT *"Boeing Tower, Cessna 1357X, unable to contact ground control, at south parking to taxi to 13L."*

PILOT *"Santa Maria Ground, Cessna 1357X, unable to contact tower, over Lompoc to land Santa Maria."*

PILOT *"Portland Clearance Delivery [from A/FD], Cessna 1357X, unable to contact ground, at Flightcraft to taxi 10L."*

Special Visual Flight Rules (SVFR)

You will need a Special VFR clearance to enter or leave the surface area of Class D airspace as depicted on your sectional chart, just as you do when Class E airspace extends to the ground. The clearance will be much easier to obtain at a tower-controlled airport than at an airport in Class E airspace simply because the communications links between towers and radar facilities are better.

If you are departing an airport where the current ceiling is less than 1,000 feet or the current visibility is less than 3 miles and you are not instrument rated, your call to ground control should be:

PILOT *"Boeing Ground, Baron 1014W, [position on the airport], request Special VFR clearance."*

You should *never* ask for such a clearance unless you are absolutely certain that the poor conditions are localized and you will encounter good VFR flight conditions before you cross the boundary of Class D airspace as shown on the sectional chart. Your clearance will sound something like this:

GROUND *"Baron 1014W is cleared out of the Boeing Delta airspace to the northwest, maintain Special VFR conditions at or below 1,300 feet. Report clear of the Delta surface area."*

When you look ahead and see the area around your destination airport is covered by low clouds, the tower controller will probably relay to your initial call with:

TOWER *"Piper 70497, Boeing Field is currently below VFR minimums, say your intentions?"*

Your reply should be:

PILOT *"497 requests Special VFR clearance into your Delta airspace,"*

and not, "I want to land."

Controllers cannot solicit SVFR; no controller will ever say "Would you like a Special VFR clearance?"

Again, you will have to read the SVFR clearance back verbatim, so be ready to copy.

TOWER *"Piper 70497 is cleared into the Boeing Field Delta airspace from the northwest, maintain Special VFR conditions at or below 1,300 feet, report entering Delta airspace."*

In case you are wondering what the "at or below 1,300 feet" is all about, the floor of Class B airspace for Seattle-Tacoma International Airport overlies Boeing Field at 1,300 feet and SeaTac is one of the airports at which no fixed-wing Special VFR is permitted in Class B airspace. You can expect similar restrictions at other airports based on local airspace allocations.

Satellite Airports in Class D Airspace

The sectional chart excerpt shows several small airports that are located within the boundaries of Class D airspace for larger airports: Hoskins (44T) at Olympia, and Spanaway (S44) at McChord AFB are examples of satellite airports. A pilot who wants to takeoff or land at a satellite airport must establish communications with the tower at the primary airport. In the case of a departure, if the tower cannot be contacted from the ground the pilot must contact the tower as soon as possible after departure.

Whether or not you need a Special VFR clearance to operate into or from a satellite airport is a function of the weather at the primary airport, not the satellite.

For example, if the Olympia weather is below basic VFR minimums, a pilot who desires to land at Hoskins (44T) must obtain a SVFR clearance to do so from Olympia Tower, even if Hoskin's weather is clear and a million. Similarly, to takeoff from Hoskins (44T) when Olympia is reporting weather below basic VFR, a pilot must communicate with Olympia Tower and get an SVFR clearance.

Arriving at Olympia

Your cross-country trip is almost over as you can see the south end of Puget Sound ahead and you know that the Olympia airport lies a few miles further south.

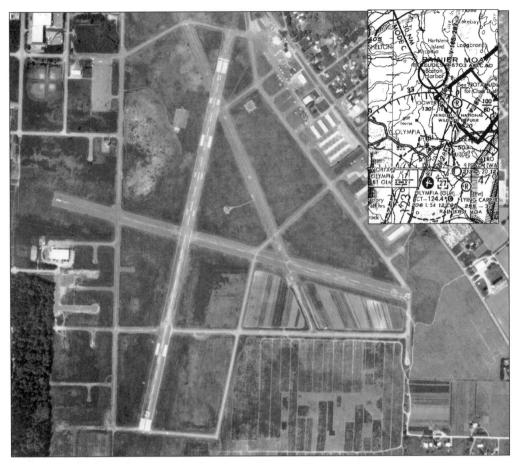

Figure 6-4. Olympia Airport, Washington

You cannot enter Olympia's Class D airspace until you have established two-way communications with the tower controller; the boundaries of the Class D airspace are printed on your chart (Figure 6-4 and airport C on the chart excerpt). You should be monitoring the tower frequency when you are no less than ten miles away from the boundary, up by Boston Harbor. When you give yourself a 10 to 15 mile buffer you can listen to tower communications with other aircraft and get an idea of what is going on. You can also hear other airplanes reporting in and get an idea of where you will fit into the inbound flow. If you are north of the airport on V-287 and hear the tower tell another airplane,

TOWER *"Runway 36 cleared for takeoff, right crosswind departure approved"*

—you know that airplane will be heading your way.

Your initial call must include your full callsign, position in relation to the airport, or a geographic reporting point. Not a point that you see over the nose, but what you see when looking straight down. Choose a reporting point that the controller will be able to recognize; even better, add something to help itinerant pilots listening on the frequency who may not be familiar with local landmarks. "Over the water tower" is not much use if there are six water towers in the vicinity. "Over the blue water tower at the southeast end of Beacon Hill" would be better.

PILOT *"Olympia Tower, Piper 70497 ten miles north for landing [or touch-and-go]"*

...should get a reply something like,

TOWER *"Piper 497, Olympia Tower, zero degree entry approved, report abeam the tower on left downwind, Olympia ceiling 6,000 broken, visibility 10, wind 340 at 6 knots, altimeter 30.11, runway 36 cleared to land."*

If you do report ten miles out, don't expect a landing clearance as a matter of course. You may be told, "Continue," followed by a landing clearance about three miles out.

This gives you permission to fly directly into the downwind leg (zero-degree entry) without making a 45. The controller could just as easily have said,

TOWER *"70497, Olympia Tower, runway 36 cleared to land. Report on the 45..."*

Have you noticed that in each of the examples the words "cleared to land" come last? That's because the FAA became aware that as soon as some pilots heard the word "cleared," their brains shut down and they didn't hear anything that followed. That is, of course, why ground controllers never use that word.

Once you are safely on the ground at Olympia you can switch over to 122.2 MHz, the standard FSS frequency, and say,

PILOT *"Seattle Radio, Piper 70497 on the ground at Olympia. Please close my flight plan at 45 past the hour."*

Once again, you have been bailed out by that Airport/Facility Directory I've been hounding you about. In the explanation of the airport listing legend, under "Communications," it says that 122.2 is assigned to most FSS's as a common enroute simplex service. The frequency 123.6 is used at nontower airports.

Strange Airport Arrival

If you are arriving at a controlled airport you are unfamiliar with, tell the controller in the first transmission:

PILOT *"Addison Tower, Baron 1014W approximately 8 miles south with information Romeo, unfamiliar."*

This will get the message across. A sharp controller will then say something like,

TOWER *"Baron 14W, do you see a brown office building at your one o'clock, five miles? Report abeam that building"*

…and when the building is off your right wing tip you will say,

PILOT *"Addison Tower, 14W abeam the building."*

The controllers have a variety of such reporting points selected, so that they can serve traffic inbound from just about any direction, and when you make that report you should be in position to at least see the runway, if not land straight in.

Including your altitude in the initial call won't do much for the tower controller, but it might be useful information for other airplanes in your vicinity. Nothing makes you scrunch up in your seat like hearing someone else report they are at your position and altitude. On the other hand, if the other pilot reports over the north end of Dog Island at 1,500 feet when you are also over the north end of Dog Island but at 1,000 feet you can notify both the tower and the other pilot of the potential conflict by transmitting:

PILOT *"Tower, 14W is also over the north end of Dog Island, level at 1,000 feet."*

The tower will then verify that the higher airplane has the lower airplane in sight and will caution both pilots to maintain visual separation and/or direct one pilot to follow the other one.

In the Pattern and on Final Approach

It is important to understand that tower controllers are not responsible for separation between VFR aircraft entering or in the pattern. That responsibility lies with the pilots, who must "see and avoid." Never rely on a controller for collision avoidance.

Many busy airports have two or more tower frequencies assigned. The ATIS will say something like,

ATIS *"…Arrivals from the northeast contact Bigtown Tower on 118.3, arrivals from the west contact Bigtown Tower on 120.2."*

A tower equipped with a BRITE radar display can see the transponder returns from VFR traffic, although the controllers cannot give vectors:

CONTROLLER *"Piper 497, suggested heading to the downwind leg is 230."*

This is not a vector. The BRITE display is a tremendous tool when the sky seems full of airplanes all headed for the downwind leg.

Tower controllers who must space traffic using binoculars or the Mark I eyeball have the problem of parallax, or difficulty in visually separating two targets along a line of sight:

TOWER *"Piper 497, turn left immediately, fly a wide downwind and follow the Baron off your right wing."*

—is a transmission that might be heard when the pilot of the Piper has the Baron in sight a half-mile ahead, while to the controller they appear about to collide. In this situation, the pilot of the airplane farthest from the pattern has a better picture of the situation than the tower controller.

PILOT *"Tower, 497 has the traffic in sight ahead and will follow that traffic."*

This is a clue to the controller that the spacing is not as tight as it appears. On the other hand, if tower instructions to enter the pattern on a 45 will put you in conflict with an airplane already on downwind you must take immediate action and then tell the tower what you are doing:

PILOT *"Tower, 497 is turning downwind outside of the traffic and will slow to follow it."*

…is one suggestion. If you wonder why your instructor has you practice transitioning from pattern entry speed to slow flight and back again, this is the explanation. Don't do any maneuvering without a clearance from the tower; the controller might have plans for the airplane following you that do not take into account the possibility that you will do an unannounced 360-degree turn for spacing. If you reduce your airspeed, the following pilot should see that the distance in trail is diminishing and follow suit.

When your position is such that you can enter the pattern on the base leg or even straight-in to the downwind (a "zero-degree" entry), don't hesitate to ask up front:

PILOT *"Tower, Cessna 1357X six miles north, request left base entry."*

—or,

PILOT *"Tower, Cessna 1357X six miles north, request straight-in to the downwind."*

In either case, the tower controller can simply turn you down:

TOWER *"Cessna 1357X, enter on the 45; report when on the 45."*

…but chances are that your request will be granted:

TOWER *"57X, cleared as requested, report two mile left base."*

This instruction means to report when your course is 90° to the extended runway centerline and you are two miles from the centerline, not when your course will intersect final two miles from the threshold. The reason for this is twofold: First, the controller wants to make visual contact and needs to know where to look; second, the controller wants to receive the report before you are close enough to mix with other traffic.

—or,

TOWER *"57X, cleared as requested, report two miles north for downwind entry."*

By the way, that "two mile left base" report means two miles from the extended centerline of the landing runway; some pilots interpret it to mean that the base leg should be adjusted so that it intersects the final approach course two miles from the runway. Look at it from the controller's point of view: if you report on base when two miles from the extended centerline, the controller knows where to look. (*See* Figure 6-5.)

A zero-degree entry to the downwind exposes you to traffic turning crosswind after departure as well as traffic entering on the 45 too close to the departure end of the pattern. It's a timesaver, but you have to keep your head on a swivel.

Always be at the published pattern altitude (in the A/FD) well before entry to avoid any high wing/low wing conflicts. Sometimes you can pick up the silhouette of other traffic against the sky, but more often than not you will be trying to pick a moving target out of a fixed background of ground structures.

Student pilots should always identify themselves as students when contacting air traffic controllers. It can't hurt, and it might help. A rapid-fire string of instructions answered with "say again, please" just won't happen if the controller realizes that slow and easy will work better.

Never be reluctant to tell a controller that you are unfamiliar with the area. Many, many pilots have blundered into the airspace of the wrong airport before confessing their confusion to ATC (and some have even landed at the wrong airport). You will receive instructions that refer to readily identifiable landmarks instead of "report crossing Washington Street."

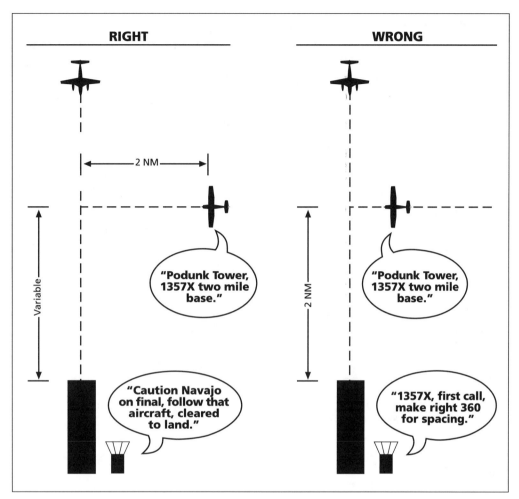

Figure 6-5. Report two mile base

Estimating distances is difficult at best. You do know how long the landing runway is, however. For example, reporting a two mile final to a 5,000-foot-long runway would require you to mentally extend the runway two runway-lengths toward your position.

If the wind is gusty and variable, don't be reluctant to ask the tower controller for a wind check on short final; remember, the ATIS might not reflect current conditions. A simple—

PILOT *"Tower, 57X, wind check."*

—will do.

Adjusting the Pattern

During your flight training, the point at which you turned from downwind to base was your own decision. At a tower-controlled airport there are a couple of situations in which that turn point might be dictated by the controller. How about:

TOWER *"Baron 1014W, follow the Cessna on final, cleared to land."*

If you don't see the airplane on final you might say,

PILOT *"Not in sight, call my base."*

But it's not always a pilot request. When a busy pattern has been stretched out to an unacceptable length, the controller might try to get things back to normal by saying to a pilot on downwind,

TOWER *"Baron 14W in sight, continue on downwind, number three to land, I'll call your base."*

In each case, simply identify yourself and say "Wilco."

At a controlled airport, is it necessary to make the base and final reports that you make at uncontrolled airports? In theory, the fact that the controller has responsibility for the airspace should relieve you of that duty. However, if you hear something on the frequency that smacks of a potential conflict, speak up — tower controllers aren't perfect. If you hear another airplane cleared to occupy the airspace that you are currently occupying, tell the tower.

At airports with parallel runways, the pilot of an airplane on a straight-in final to one runway (completing an instrument approach, for example), might be concerned to see you in position to cross its path.

PILOT *"Baron 14W turning final for 13L has the traffic on final for the right"*

…will lower that pilot's anxiety level.

You're the boss when it comes to the operation of your airplane. That's difficult for low-time pilots to accept, but it is right there in the Federal Aviation Regulations. You must add "negative" and "unable" to your vocabulary. Abusing the brakes to comply with an instruction to turn at the next intersection is potentially much more expensive than saying "unable," because on a slick runway you might lose control of the airplane. Keep in mind that the controller does not have to pay for any damage you do to your airplane and will go home with a clear conscience.

This is equally true when you get clearances like "Make short approach," "Cleared to land runway 36, hold short of runway 27," and things like that. Don't make extravagant efforts to comply with controller instructions that will stretch your abilities, and the capabilities of your airplane, to their limits.

If you are given a "land and hold short (of a crossing runway)" instruction and there is any doubt whatsoever about your ability to get your airplane on the ground and stopped before reaching the crossing runway, let the controller know as soon as possible so that airplanes using the crossing runway can be spaced accordingly. This is not a situation in which you want to surprise the controller.

When appropriate, suggest alternative actions. If runway 36 is the active, but runway 30 will get you to the ramp with less taxi time, ask the controller for clearance to land on 30 instead.

PILOT *"Tower, Cessna 1357X requests 30"*

—will do the trick if the controller thinks that no compromise to safety will result. The same holds true for departures. Traffic and wind permitting, you can take off on any runway you choose—a tower controller cannot deny you a takeoff clearance. Of course, you might have to sit and stew for a long time while everyone else takes off on the preferred runway, but your time will come if you are patient.

Every year, FAA air traffic controllers at Wittman Field in Oshkosh, Wisconsin, turn its 8,000-foot-long runway into two end-to-end 4,000-foot runways. One pilot will be cleared to land on the numbers, while the next will be told to maintain altitude while flying down the first half and to land past the midpoint. No one taxis to an intersection, they just turn off into the grass and are directed to parking. For one week, Oshkosh becomes the busiest airport in the world, with over 15,000 airplanes coming in to land.

When you hear a tower controller say something like,

TOWER *"57X, make midfield base south of the tower, cleared to land on the last half,"*

(or, when you are on final)—

TOWER *"Keep your altitude up, land on the last half of the runway"*

…it's not something unusual, but is a tested procedure that can expedite your landing.

Check to see if your local pilot supply store has any audio tapes of tower communications at Oshkosh—listening to that activity will really open your eyes to how flexible aviation communication can be.

"Land long" is another phrase that can be used by either the pilot or the controller. Your airplane would rather fly than taxi, so why not fly down the runway to shorten taxi time when that is convenient?

PILOT *"Tower, 57X requests long landing"*

…is all it takes. A clearance to land long is not a sure thing—hearing,

TOWER *"57X, unable, traffic waiting for midfield departure"*

—is one possibility. If your home airport has a tower and the controller recognizes your N-number, you may hear,

TOWER *"Cessna 57X, where do you park?"*

followed by (if it would save you some taxi time),

TOWER *"Cleared to land, long landing approved"*

—(although you didn't ask for one).

When landing at an airport with parallel runways, don't be reluctant to turn from base to final just a little early to insure that your flight path will not encroach on the other runway's final approach path. This is a situation where those crisp 90-degree turns are not appropriate.

A tower controller may ask you to slow down, speed up, do S-turns for spacing, or even do a 360-degree turn for spacing on downwind or final. That's fine—do as requested if you can do so safely (or say "unable"). Do not do any maneuvers on your own authority just because it seems like the right thing to do; the controller has a landing sequence in mind and your well-intended maneuver might just throw a wrench into the works.

At a busy airport you may hear,

TOWER *"57X, cancel landing clearance, go around at pattern altitude."*

Unexpected, but not a big deal—just make your rectangular pattern at pattern altitude with the upwind leg right over the runway while the tower gets a couple of departures out.

PILOT *"57X, Wilco."*

If you are on final and getting too close to the airplane ahead, the tower controller may ask you to go around without overflying the other plane:

TOWER *"57X, go around, overfly the west taxiway [or the grass]."*

Again, you simply acknowledge with your callsign. "Wilco" is the official AIM response, but when the tower controller can see that you are complying it is superfluous.

"Direct to the numbers" is something you might hear when a controller wants to speed things up by having you skip the base and final legs. When you hear it, just point the nose of the airplane at the numbers (actually, somewhat short of the threshold if you want to avoid floating forever) and continue your descent.

Night Operations

A night landing at any airport can be a challenge, but if you know the magic words, a tower controller just might be able to make it easier. The instrument approach lights are normally turned on only when the ground visibility drops below two miles, but the controller will be happy to turn them on if you say something like,

> **PILOT** *"Bigtown Tower, request the approach lights and rabbit."*

The "rabbit" is a system of sequenced flashing lights that can't be mistaken for anything else and it leads directly to the threshold (this works in the daytime, too). Runway End Identifier Lights (REIL) are two lights placed on either side of an instrument runway that flash regularly; if the runway is not served by a rabbit, the REIL will do the trick. The controller can also vary the intensity of the runway lights on request.

As you know, not all tower-controlled airports are 24-hour facilities. You should be able to control the approach and runway lights by keying your microphone the correct number of times on the correct frequency. Find the details on a specific airport in the Airport/Facility Directory, a reference you should not be without.

Landing Alternatives

Most of your landings at controlled airports will be full stops or touch-and-goes. Make sure that the controller knows what your needs are well in advance. If you intend to land and taxi back, or just quit for the day, and the controller says,

> **TOWER** *"Cleared for touch and go"*

…be sure to respond with,

> **PILOT** *"1357X is full stop."*

Although the alternatives I am about to suggest are more for instructors than for students or certified pilots, you should be familiar with them.

The first is "stop and go." When practicing landings, you might want to bring the airplane to a full stop, configure for takeoff, and then commence your takeoff.

You clue the controller by asking for a stop-and-go, so that spacing between your airplane and following airplanes can be adjusted.

The next is "cleared for the option," and you get it by saying:

PILOT *"Tower, 1357X requests the option."*

When you are cleared for the option you can make a touch-and-go, a stop-and-go, or a full stop; the controller will give you time and space for any of the above.

Land and Hold Short (LAHSO)

If you are inbound to an airport that has intersecting runways and the tower controller says,

TOWER *"...cleared to land runway 21 left, hold short of runway 30"*

you must consider your speed, altitude, aircraft configuration, and proficiency, and if you are not absolutely certain that you will be able to bring your airplane to a controlled stop before reaching the intersection you must say "Unable." That will give the controller an opportunity to change plans and give you alternative instructions. Smoking the brakes as you skid through the intersection is definitely bad form. You should receive LAHSO instructions only when the reported weather meets basic VFR minima: a ceiling of at least 1,000 feet and light visibility at least 3 statute miles.

PILOT *"Prescott Tower, Cessna 1357X, what is my landing distance on 21 Left?"*

TOWER *"Five thousand, four hundred feet."*

PILOT *"Unable." Or "1357X is a student pilot."*

TOWER *"Cessna 1357X, go around at pattern altitude, make right traffic for runway 30. Citation 43G cleared for takeoff runway 30."*

The Special Notices section of the *Airport/Facility Directory* lists controlled airports where LAHSO operations may be required, together with the available landing distance (ALD) for runways involved. This determination should be part of your preflight planning process. Tower controllers will provide ALD information on request, and it should be included in the ATIS broadcast.

It is your responsibility to reject a LAHSO clearance if you are uncertain of your ability to bring your airplane to a stop in the available landing distance. Student pilots *should not* accept a LAHSO clearance.

NORDO

Operating at a tower-controlled airport without a radio is legal, but it places an undue load on the controllers and should be avoided if possible. Handheld radios aren't that expensive. This book is intended for pilots with radios installed in their airplanes.

What if your radio breaks down in flight? The smart thing to do is to land at an uncontrolled airport and call your destination tower on the telephone, giving your airplane type, color, and your estimated time of arrival. Then just fly into the pattern, blinking your landing light and waggling your wings until you get a green light from the tower. You could do the flashing-lights and wing-waggle bit without the phone call, but it isn't good form to do so. (*See* Chapter 12 for more about emergency procedures.)

Departing a Satellite Airport

When you are departing from a small airport that lies within the boundaries of Class D airspace, you are required to contact the control tower at the primary airport from the ground before takeoff if that is possible:

> **PILOT** *"Bigtown Tower, Piper 70497 ready for takeoff at Littlefield, VFR westbound."*

If you cannot contact the tower from the ground, you must do so as soon as possible after departure:

> **PILOT** *"Bigtown Tower, Piper 70497 off from Littlefield, through 500 climbing 2,500, VFR westbound."*

In either case, the tower will acknowledge your call and give you traffic information.

If your destination airport lies within the boundaries of Class D airspace, you must establish communication with the tower just as if you were going to land at the primary airport. The only exception is that you say:

> **PILOT** *"Bigtown Tower, Piper 70497 six miles southeast with information Mike, landing Littlefield."*

The tower will simply acknowledge your call and give you any pertinent traffic information. Do not expect to receive a clearance to land—the tower controller does not have any clearance authority at the satellite airport.

Just Passing Through

If you want to fly through Class D airspace there is no reason why you shouldn't.

Just give the tower a call a few miles out:

> **PILOT** *"Bigtown Tower, Cessna 1357X five miles south, request transition south to north at 1,800 feet."*

> **CONTROLLER** *"Approved as requested. Report clear."*

Remember that Delta airspace extends from the surface to 2,500 feet above airport elevation (which isn't necessarily the surface — *see* Page 6-2) unless the top 500 feet has been released to the overlying radar facility. In that case, if you are receiving radar flight following services you can breeze right on through. However, if you are any lower than 2,000 feet above the field and receiving flight following, it wouldn't hurt to check with the controller:

> **PILOT** *"Approach [or Center], 1357X, am I cleared through Bigtown's Delta airspace?"*

(The AIM and the *Air Traffic Control Handbook* differ slightly on this, so call to be on the safe side.) Monitoring the tower frequency would be a great idea, however. You might hear something like:

> **TOWER** *"Baron 14W is cleared for takeoff; there is a Cessna one mile south of the airport, eastbound, indicating 3,000 feet, but we're not talking to him."*

Or, you might hear,

> **TOWER** *"Cessna 1357X there is a Learjet at your two o'clock on final for runway 6."*

That would be your clue to keep an eye on the Learjet until it was well clear of your flightpath, or you could volunteer:

> **PILOT** *"Tower, Cessna 1357X has the Learjet visually and will pass well clear."*

The best place to cross an airport is overhead and at right angles to the threshold of the active runway. Planes landing or taking off will be either on the runway or very close to it at that point. If your desired penetration altitude will be within 2,500 feet of the surface, just call the tower well outside of the Class D airspace and say,

> **PILOT** *"Bigtown Tower, Cessna 1357X request clearance through your airspace at [your desired MSL altitude],"*

…and the tower will reply with something like,

> **TOWER** *"57X cleared as requested, cross the upwind end of [active runway] at [assigned altitude] and report when clear."*

Some busy airports have VFR flyways designed to enable VFR flights to get around or through the tower's airspace with a minimum of disruption; these flyways are printed on the reverse of the Terminal Area Charts for the airports involved. Pilots must still communicate with the tower (or Approach Control — the Terminal Area Chart will have the details), but at least they know that there is a route reserved for VFR operations and that their chances of getting across the airport are much better using the flyway than by using a "roll-your-own" route.

Los Angeles International is a special case. It not only has *two* flyways, but it also has a "Special Flight Rules Area." Pilots using the SFRA are not required to communicate with ATC and do not require a clearance of any kind through the Class B airspace — but they must have a current Terminal Area Chart in the cockpit because that is where all of the operational details will be found. A discrete frequency is assigned solely for the use of pilots transiting the SFRA, so that they can talk to one another and avoid any conflicts in this very small corridor.

When Your Eyes Deceive You

Occasionally, the FAA will establish a temporary control tower at an airport in Class G or E airspace for an airshow or a fly-in. The existence of these towers is promulgated through Notices and Letters to Airmen, but there is obviously no change to the relevant sectional chart. Still, when a temporary tower is in operation it brings Class D airspace with it. If you fail to communicate with the tower before landing at such an airport and get a violation, "It wasn't on the chart" will not be accepted as a defense.

You might find a permanent tower in Class G airspace, indicated by magenta tint (the floor of Class E airspace is 700 feet AGL) but with a blue airport symbol; Lake City, Florida and Trent Lott, Alabama are two examples. In those situations, Class G cloud clearance and visibility requirements must be observed and you still have the communication requirement that exists at any towered airport.

UNICOM at Tower-Controlled Airports

Almost all of the tower-controlled airports on the sectional chart excerpt show a frequency of 122.95, which is a UNICOM frequency. Does this mean you can bypass the tower controller and ask UNICOM for the runway in use? No way. UNICOM is provided at tower-controlled airports for many of the same reasons it is provided at small airports: calling for fuel, calling a cab, or notifying someone of your ETA, but not for traffic or weather information. Most UNICOM operators at tower-controlled airports can't see the runway or the traffic pattern from their offices and have no weather reporting instruments, so you won't be using UNICOM for anything other than the types of transmissions listed above.

At some airports with part-time towers, you will find both a UNICOM frequency and a CTAF. Keeping in mind the limitations on the use of UNICOM at an airport with a tower, always use the CTAF for advisories and position reports. Your destination FBO may have its own frequency (*see* Page 4-3).

Summary

When the tower is in operation, two-way radio communication is required in Class D airspace (or in Class E or even Class G if a temporary tower has been established by NOTAM). When the tower is closed, the airspace becomes either Class E (full-time weather observer) or Class G. In less than VFR conditions, operation in Class D or E airspace requires an SVFR clearance. Inbound, call the local controller well in advance of crossing the blue dashed line.

Radar Required

The FAA designates airports as Class C airspace when radar control is available and the airspace is congested (but not sufficiently congested to warrant Class B status). *See* Figure 7-1 on the next page.

Before you can take off from or land at the primary airport or fly through the Class C airspace, you must establish two-way communications with the controlling facility.

When Class C is Not Class C

All Class C airspace is not created equal—in some locations it is a 24-hour-a-day operation, while in others the Class C is in effect part-time. If your sectional chart simply says "Class C," near the magenta circles, it is in effect full-time. If it says "See NOTAMS/Directory for Class E (or G) sfc eff hrs," it is part-time. Look at the chart excerpts for Fairchild AFB and Burlington Airport (airport K) on the foldout panel; Fairchild depicts full-time, Burlington depicts part-time. Also, look at Figure 7-2 to see how this information is presented on the back of the sectional.

You will notice on the chart excerpt that there is no blue dashed line around Spokane International Airport (airport H on the chart excerpt) and the adjacent Fairchild AFB, although both have towers. The Class C airspace is effective full-time at those airports, and because the terminal radar facility works hand-in-glove with the control tower, there is no Class D airspace as such.

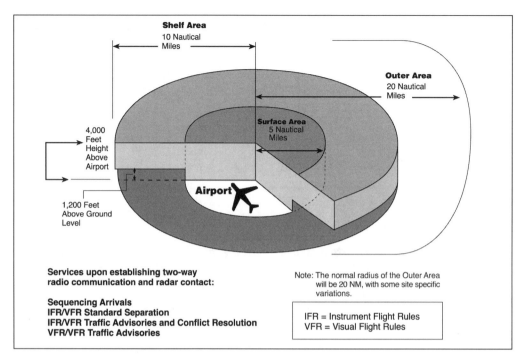

Figure 7-1. Class C airspace

CONTROL TOWER	OPERATES	TWR FREQ	GND CON	ATIS	ASR/PAR
BANGOR INTL	CONTINUOUS	120.7 257.8	121.9 348.6	127.75	ASR
BURLINGTON INTL	0500-0100	118.3 257.8	121.9 348.6	123.8 269.9	ASR
MONTREAL INTL (DORVAL)	CONTINUOUS	119.9 267.1	121.9 275.8	133.7 127.5	
MONTREAL INTL (MIRABEL)	CONTINUOUS	119.1 282.4	121.8 246.6	125.7 126.1	
OTTAWA/MACDONALD-CARTIER INTL	CONTINUOUS	118.8 236.6	121.9 275.8	121.15 265.6 132.95 382.05	
QUEBEC/LESAGE INTL	CONTINUOUS	120.3 236.6	121.9 250.0	134.6 128.3	
ST-HUBERT	0600-2400 APR-OCT 0600-2300 NOV-MAR	118.4 352.5	126.4 283.4	124.9 124.1	
ST-JEAN	0730-2130 APR-OCT 0800-2100 NOV-MAR	118.2 356.8	121.7		
VALCARTIER DND	LIMITED HOURS MON-FRI	126.2 307.6			
WHEELER-SACK AAF	CONTINUOUS	118.75 241.0	121.9 229.8		ASR/PAR

CLASS B, CLASS C AND SELECTED RADAR APPROACH CONTROL FREQUENCIES

FACILITY	FREQUENCIES	SERVICE AVAILABILITY
BANGOR CLASS C	124.5 239.3 (335°-154°) 125.3 239.3 (155°-334°)	CONTINUOUS
BURLINGTON CLASS C	121.1 396.1 O/T 120.35 380.3 ZBW CNTR	0500-0100 O/T CLASS E
WHEELER-SACK AAF RADAR	128.25 299.85	CONTINUOUS

**ZBW-Boston
O/T indicates Other times

Figure 7-2. Service availability

Transponder Use

Because Class C airspace is designated where a terminal radar is available, pilots must have an operable transponder with Modes A and C in their aircraft for identification.

More often than not, ATC controllers will assign a discrete transponder code for use while in their airspace. Many instructors tell their students that it is not necessary to acknowledge ATC instructions to change transponder codes or IDENT because the controller will see the change on the radar scope. However, controllers are quick to point out that they have other things to do in addition to watching for your data block, and several seconds may elapse before the change shows up on the scope. Moral: Always back up transponder changes with a voice transmission —

> **CONTROLLER** *"Piper 70497, squawk 6723 and ident."*

> **PILOT** *"497 squawking 6723"*

—will do the trick.

Arrival

The "outer area" of Class C airspace extends 20 nautical miles from the primary airport and affords you the opportunity to make contact with the controlling ATC facility well in advance of entry into the shelf area, 10 miles in radius, and from 1,200 to 4,000 feet above field elevation, where participation is mandatory. Take advantage of it — call approach control 15 to 20 miles out so that you can devote all of your attention to flying the airplane and checking for traffic. Bear in mind that you cannot enter the 10-mile radius shelf area without having established communication with ATC. The Class C airspace at Spokane International Airport (Figure 7-3 and airport H on the chart excerpt) is a good example; you get the frequencies from a panel on the Sectional chart or from the A/FD:

> **PILOT** *"Spokane Approach, Baron 1014W 20 miles west, 7,500 feet with November, landing Spokane International."*

> **APPROACH** *"Baron 14W, Spokane Approach, maintain present heading, descend and maintain 6,000 feet, squawk 4657."*

> **PILOT** *"Leaving 7500 for six thousand, 4657, 14W."*

Although you do not need a clearance in order to enter Class C airspace, when Approach Control has established radar contact and you are talking to each other,

Continued on Page 7-6

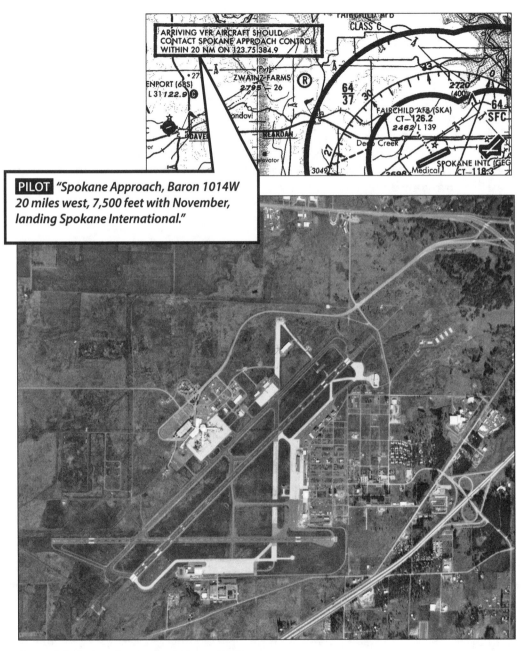

PILOT *"Spokane Approach, Baron 1014W 20 miles west, 7,500 feet with November, landing Spokane International."*

Figure 7-3. Spokane International Airport, Washington

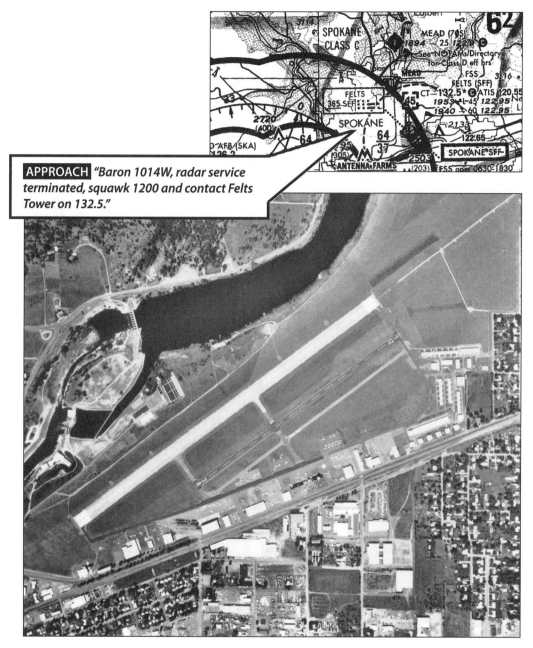

APPROACH *"Baron 1014W, radar service terminated, squawk 1200 and contact Felts Tower on 132.5."*

Figure 7-4. Felts Field, Spokane, Washington

you must comply with any heading or altitude changes directed by Approach. This is a real advantage to a pilot who is a stranger to the area, because the controller can provide vectors to the pattern or to the final approach course. After coordinating with the control tower, the approach controller will give you the tower frequency and tell you to contact the tower. What could be easier?

Essentially the same conversation would take place if Baron 14W intended to land at Felts Field, outside of the 5-mile radius surface area (Figure 7-4 and airport I on the chart excerpt). In this case, the original call would be,

> **PILOT** *"Spokane Approach, Baron 1014W, 20 miles west, 7500 feet, landing Felts."*

Spokane Approach would hand off 14W to the Felts tower as soon as it was clear of the surface area and advise the pilot to contact the tower:

> **APPROACH** *"Baron 1014W, radar service terminated, squawk 1200 and contact Felts Tower on 132.5."*

In addition to staying on the right side of the law, being in contact with ATC in Class C airspace can really help when you approach a strange airport; just ask for vectors to the pattern.

> **PILOT** *"Bigtown Approach, Baron 14W is unfamiliar, request vectors to the pattern."*

More often than not, this will result in your being placed in position for a straight-in approach, but will on occasion place you on downwind. The last transmission you will hear from the radar controller will be,

> **APPROACH** *"Baron 14W, contact tower 118.1."*

The tower controller will have been alerted by interphone and will know where to look for you when you make your call:

> **PILOT** *"Bigtown Tower, Baron 14W on the 45 for 13 left."*

Departure

When you depart the primary airport in Class C airspace you are required to establish two-way communication with ATC, and your normal contacts with ground control and the tower meet this requirement (but if there is a clearance delivery frequency in the A/FD, use it). You may receive specific instructions from the tower—

> **TOWER** *"After takeoff turn right heading 030, climb and maintain 2,500 feet"*

—to keep your flight separated from other departures.

When you are departing Class C airspace, give ATC a call upon leaving the shelf area (10 miles out):

PILOT *"14W requests terminate radar service"*

…which will bring,

CONTROLLER *"Baron 14W, position 11 miles southeast of Bigtown, radar service terminated, squawk one two zero zero. Frequency change approved."*

If you do not make this call, the radar facility must continue to provide traffic vectors or traffic advisories until your radar target leaves the outer area (20 miles out). ATC's final transmission will be the same.

SVFR

If the ceiling is below 1,000 feet and/or the visibility is less than 3 miles, you must either go IFR, or ask ATC for a Special VFR clearance, just as in Class D airspace. To accept such a clearance, the visibility must be at least one mile and you must be able to stay clear of clouds while within the surface area of the Class C airspace. I've hedged on identifying the type of visibility because it depends on whether or not there is an observer at the airport to report ground visibility. If there is, the ground visibility must be at least one mile, and if there isn't, then flight visibility governs.

Satellite Airports

When departing a satellite airport in Class C airspace you must contact ATC as soon as practicable after takeoff:

PILOT *"Bigtown Approach, Piper 70497 just off from Littleburg, 1,500 climbing 3,500."*

You will be asked to either IDENT on code 1200 or change to a discrete code, then:

APPROACH *"497, radar contact, report level 3,500."*

If you intend to land at a satellite airport inside Class C airspace, simply advise ATC on initial contact:

PILOT *"Bigtown Approach, Piper 70497 15 miles northwest, landing Littleburg."*

Remember that a column of airspace extending from airport elevation to 10,000 feet MSL and 20 miles in diameter extends above the primary airport in every Class C airspace. The only direct impact this has on your operation is that you must have a Mode C encoding transponder when transiting this column of airspace. The bottom 4,000 feet of this column of air (or as charted) is the Class C airspace where two-way communication with ATC is required.

"Piper 70497, Stand By"

There is a loophole in Class C procedures that I am honor-bound to tell you about, although I personally would not take advantage of it. If you are inbound to a Class C airport and say,

> PILOT *"Podunk Approach, Piper 70497, ten miles northeast, landing"*

…and the controller says,

> APPROACH *"Aircraft calling Podunk Approach, stand by"*

—you have no choice but to stay outside of the Class C airspace until the controller has a chance to say,

> APPROACH *"Aircraft calling Podunk Approach, say again your callsign."*

You will provide your callsign, of course, and everything will then proceed normally. If, however, the controller answers your initial call with,

> APPROACH *"Piper 70497, stand by"*

…the AIM says that communications have been established (through the use of your callsign) and you can proceed into the Class C airspace. I wouldn't touch that one with a ten-foot pole. If the controller is too busy to handle your flight normally, having you barge into the Class C airspace is ludicrous. Even though the Federal Aviation Regulations let pilots do all sorts of things, many of those things may be marginally safe—but stupid. This is an example.

TRSAs

Terminal Radar Service Areas (TRSAs) are on a par with dinosaur footprints—they are reminiscent of days gone by. Usually associated with military or joint-use airports, the existence of a TRSA tells you that radar services are available but not mandatory, and that the airport is not busy enough to warrant its designation as Class C airspace. TRSAs show up on sectional charts as black circles.

If your destination is an airport within a TRSA I strongly recommend that you take advantage of the services being offered. Forget whatever you have read about Class II or Class III service—those distinctions no longer exist. You will find the radar approach control frequency on your sectional chart. Dial it up and say,

> PILOT *"Yeager Approach, Baron 1014W is 12 miles east, landing Yeager."*

The radar controller will assign a discrete squawk, and acknowledge radar contact. You will be provided with traffic information and vectors to the pattern and then handed off to the tower controller. When directed to change to the tower, remember that the tower and radar controllers have been discussing you on the telephone and the tower controller know all about you.

PILOT *"Yeager Tower, Baron 1014W on left base [downwind, final] for runway 27."*

TOWER *"14W cleared to land [or follow the Belchfire on final]."*

Departing an airport in a TRSA, you should once again use the radar services that are available to you. Use normal Class D airspace departure procedures, and when airborne, ask the tower for a frequency change to the radar control frequency; then ask the radar controller for VFR traffic advisories while leaving the TRSA. Respond to traffic advisories with "In sight" or "Negative contact, keep me advised." Do not use "no joy" or "tally ho" unless you flew for the RAF in World War II.

The radar controller will turn you loose by saying "Radar service terminated, squawk 1200, frequency change approved." Don't leave the TRSA frequency without asking the controller for the local Center frequency, because you just might be able to switch over to Center for radar flight following (*see* Chapter 5, Class E airspace).

PILOT *"Metro Center, Baron 1014W is 8 miles northeast of Yeager, VFR at 4,500 feet. Request flight following to Bigburg."*

CENTER *"14W, Metro Center, radar contact, maintain VFR, advise any altitude changes."*

If you want to simply fly through TRSA airspace you are free to do so without contacting anyone. However, if the ceiling permits flight under VFR but still holds you down to a relatively low altitude where you might come into conflict with traffic entering or departing the TRSA, a call to clarify your status would eliminate any doubt for the controller:

PILOT *"Yeager Approach, Baron 1014W is 8 miles west, will be flying through your airspace at 4,500 feet west to east."*

CONTROLLER *"14W, Yeager Approach, squawk 0400, report clear."*

Now the controller knows who you are, what you are up to, and does not have to worry about an unknown target suddenly mixing it up with the TRSA traffic. You will be sent back to code 1200 when you report clear of the TRSA.

Summary

Entry into Class C airspace requires that two-way radio communications be established between the pilot and the controlling radar facility. A transponder with altitude reporting capability is required. The outer area gives an inbound pilot plenty of time to get his or her act together before entering the airspace. Pilots operating to or from satellite airports must contact the controlling facility as soon as possible.

Clearance Required

This is the first discussion of an airspace classification that requires a specific clearance before any operation within its boundaries. Class B airspace exists at the very busiest airports in the country, and the FAA exercises strict control over operations in that airspace. It even has its own chart series, the Terminal Area Charts (TAC). Student pilots are not allowed to fly in Class B airspace without specific instruction and an instructor's endorsement in their logbooks. We'll assume for the purposes of this discussion that you hold at least a Private Pilot certificate as you fly toward a destination that is the primary airport for Class B airspace (a properly checked-out student will follow the same procedures, of course).

Many pilots who are uncertain of their ability to fly in congested airspace will fly many miles out of their way to avoid flight in Class B airspace. In my opinion, that is a mistake. You have the right to exercise the privileges of your certificate and should assert that right. Of course, if the ATC controllers workload is such that handling VFR flights would interfere with ATC's basic responsibility to handle IFR flights, they have the right to say "unable due to workload." Then you must fly the long way around, and you should include this possibility in your preflight planning.

If you are forced to circumnavigate Class B airspace, try to stay at least one mile away from its boundaries and at least 1,000 feet beneath its floor. The FAA determines the boundaries horizontally but distance measuring equipment (DME) measures slant range. When your DME reads 10.0 miles, you may be only 9.6 horizontal miles away from the primary airport (depending on altitude, of course), and violating the airspace boundary.

Don't try to fly over Class B airspace unless you can clear it by 2,000 feet or more — turbojets like to pop out of the top if they can climb rapidly enough.

You should have the appropriate Terminal Area Chart for a couple of reasons. First, the ATC frequencies are listed in more detail, and second, the charts show VFR reporting points that controllers want you to use.

Approach and Departure Control

The controllers you will be talking to in Class B airspace occupy hushed, almost totally darkened rooms near the control tower for the primary airport. The walls are lined with radar scopes, with a team of controllers at each position. The airspace is divided into sectors, and in most cases is divided vertically as well, so that the same geographic area may be covered by a high sector controller and a low sector controller. Each controller has several frequencies available (some are Ultra-High Frequencies for the military). *See* Figure 8-1. You can see how frequency changes might be required as you cross sector boundaries. The circles are drawn around high obstacles or terrain to ensure that you are not vectored into something hard.

You will find the Automated Terminal Information Service frequency for the primary airport on the chart (some airports have more than one ATIS). Try to copy the ATIS from as far away as possible so you know which runways are in use; that may affect the frequency on which you make your initial contact. Select the ATIS frequency and break the squelch to see if you can pick the information out of the static. If you can't, try again in a few minutes. Waiting until the ATIS signal is strong enough to break the squelch deprives you of planning time. The ATIS broadcast might include the initial contact frequency, but you will be ahead of the game if you have checked the Airport/Facility Directory and the TAC.

Just Passing Through

Let's fly through the Class B airspace first. You've looked up the ATIS frequency and listened to the taped broadcast, now ask yourself what information a controller might need from a pilot who wants to fly through Class B airspace: Who, Where (including altitude), and What (intentions).

PILOT *"Metro Approach, Baron 1014W over the BIGTOWN VOR at 7,500 feet, request clearance eastbound through your Bravo airspace."*

That transmission assumed that you listened on the approach control frequency and found it fairly quiet. If it was hard to get a word in edgewise you could try,

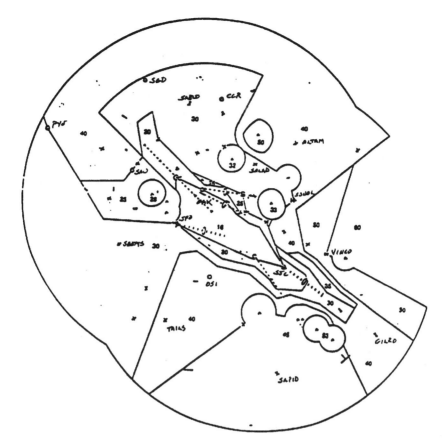

Figure 8-1. Bay approach sector chart (San Francisco, California)

PILOT *"Metro Approach, Baron 1014W, request."*

That alerts the controller you are on the frequency and want something. Your tail number will be added to the list of folks who want something, and the controller will get back to you as soon as practicable.

Your initial transmission didn't include a transponder squawk for a very good reason. If you were squawking anything other than 1200 it would be a discrete code assigned by ATC (an en route Center, most likely), and the whole handoff procedure would be handled between Center and Approach on the telephone. Center would tell you to contact Approach (and give you the correct frequency) and all you would have to do is switch frequencies and say,

PILOT *"Metro Approach, Baron 1014W, level at 7,500 VFR."*

Your position, altitude, and intentions would have been passed along by the Center, but every controller wants to hear the altitude directly from you. Reminding the controller you are VFR keeps that busy individual from riffling through a pile of flight strips looking for your IFR clearance. From this point on, passing through is just like landing at the primary airport; you just won't be turned over to the tower for landing instructions.

Landing at the Primary Airport

If you plan to land at the primary airport, make this clear in your initial transmission:

> **PILOT** *"Bigtown Approach, Piper 70497 10 miles northwest [or "over the freeway interchange"] with Kilo, landing Bigtown."*

Try to use one of the charted VFR reporting points. Do not get closer than one mile to the boundary of the Class B airspace (two miles is much better) until you hear the words,

> **APPROACH** *"Piper 70497, radar contact, cleared to operate in the Bigtown Bravo airspace, maintain VFR."*

"Radar contact" alone is not sufficient. Your response should be:

> **PILOT** *"Understand cleared into the Bravo airspace, 497."*

If you are uncertain, ask:

> **PILOT** *"Is Piper 497 cleared into the Bravo airspace?"*

Get your voice on the ATC tape recorder and make no assumptions—it is your pilot certificate on the line, not the controller's.

The boundaries of the Class B airspace that the controller sees on the scope are not as precise as what you can see by looking at objects on the surface, but if the controller decides that you have entered the airspace without a clearance, the government's computer record of radar images will prevail.

The Approach controller may ask you to squawk IDENT if you are on code 1200, assign a discrete code while in their airspace, or simply leave you on the code assigned by Center.

If you intend to land at the primary airport or at a satellite airport located within the Class B surface area, say,

> **PILOT** *"Metro Approach, Baron 1014W over the BIGTOWN VOR at 7,500 feet, landing at Metroplex International [or at Suburbia Municipal]."*

Expect the controller to hand you off to the Metroplex International tower controller with no further action on your part:

APPROACH *"Baron 1014W, contact Metro Tower 120.6"*

PILOT *"Roger, 120.6, 14W"*

—or to vector you to the Suburbia pattern and turn you loose as soon as you report the field in sight:

APPROACH *"Baron 1014W, squawk 1200, change to advisory frequency approved; good day."*

Ideally, you should pick a VFR reporting point designated by a little magenta pennant on the Terminal Area Chart, or an actual photograph of a ground feature:

PILOT *"Metro Approach, Baron 1014W over the stadium at 7,500 feet..."*

...or give a direction and distance from the primary airport:

PILOT *"Metro Approach, Baron 1014W, 18 miles northwest at 7,500 feet..."*

If you are asked to IDENT, do so and confirm it by voice; the ATC radar antenna turns slowly enough that several seconds might go by before the controller notices your IDENT.

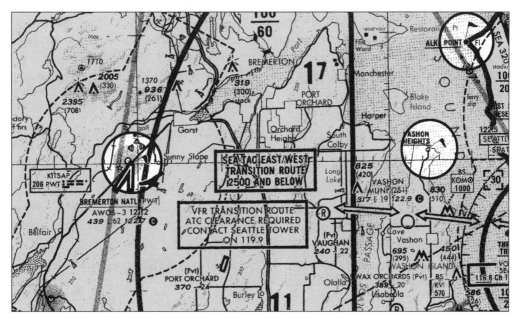

Figure 8-2. Terminal Area Chart, VFR reporting points

While operating in Class B airspace, either landing or passing through, the controller can assign headings and altitudes that will keep you clear of IFR traffic and you must comply; however, if compliance would take you into a cloud you *must* say,

> **PILOT** *"Unable to maintain VFR on that heading [at that altitude], 14W."*

…and the controller will provide alternate instructions. No controller can authorize you to violate the VFR minimums contained in Part 91.

Instrument pilots have to read back to the controller any clearances that include changes in heading or altitude. There is nothing in the regulations that says VFR pilots have to do the same thing, but it makes such good sense that there is no reason not to.

> **PILOT** *"Baron 14W leaving 7,500, descending 4,000"*

> **PILOT** *"Baron 14W, left to 340."*

Once you have been handed off to the control tower operator there is no difference between an airport in Class B airspace and any other controlled airport.

Departing an Airport in the Class B Surface Area

The primary airport in a Class B surface area is a big time operation. There were over 30 such areas in 2004 with more under consideration. These are airports with heavy concentrations of air carrier airplanes and you have to be on your toes. You don't have to be super-pilot, but you do have to stay alert and respond to instructions quickly and accurately without spending too much time talking on the radio.

Everything in the discussion of Class D airspace applies, because this is basically a tower-controlled airport with a few extra requirements. Once again, you should have a Terminal Area Chart on board — if there are any specific requirements they will be on that chart in the form of notes.

If the TAC or the Airport/Facility Directory lists a Clearance Delivery frequency, call that frequency first (instructions to do so might be on the ATIS). Just select the Clearance Delivery frequency and say,

> **PILOT** *"Metro Clearance, Navion 5327K at the south ramp, VFR to Bigburg. Request clearance to operate in Bravo airspace."*

The CD controller will give you a clearance that will get you on your way and then tell you to contact Ground Control:

`CLEARANCE DELIVERY` *"Navion 5327K is cleared into the Class B airspace via fly runway heading, maintain VFR at or below 1,500, expect 4,500 ten minutes after departure. Departure frequency 125.05, squawk 0203. Contact Ground on 121.9 when ready to taxi."*

Now you are in the same situation as an instrument pilot; you must read the clearance back to the controller to ensure understanding. The fact that you are VFR will let the controller know copying clearances might be a new experience for you. When you call Ground Control for taxi instructions, let that controller know that you have departure instructions:

`PILOT` *"Metro Ground, Navion 5327K on the south ramp, VFR to Bigburg with departure instructions."*

If the airport does not have a Clearance Delivery frequency, call Ground Control as you would at any tower-controlled airport.

Pre-Taxi clearance (Cpt) is a special situation designed for instrument flights only. If the Airport/Facility Directory lists both Clearance Delivery and Pre-Taxi Clearance Delivery, do not use the pre-taxi frequency.

Everyone in the tower knows that you need a clearance to operate in the airspace, but you should not assume that a clearance will come automatically. Don't even accept a "Position and hold" clearance until you know that operation in Class B airspace has been approved. You will get a discrete transponder code to squawk, don't cross the hold line without one.

Once again, expect headings and altitudes that will keep you clear of IFR flights until you have crossed the Class B airspace boundary. The tower might say something like,

`TOWER` *"Navion 5327K, suggest heading 250 [or maintain runway heading], maintain VFR, contact Metro Departure on 125.05."*

Don't switch your transponder to the ON position until you are on the runway and in position for takeoff. When the Departure controller can see your transponder return while you are still on the ground, he or she will have accepted responsibility for you before the tower instructs you to change frequencies. If you have to gain a little altitude before Departure can see your return, the "Contact Departure" instruction will be delayed until coordination has been accomplished. Don't change frequencies on your own; if it seems as though you have been forgotten,

`PILOT` *"Do you want 27K to go to Departure?"*

...should elicit a frequency change from the tower controller.

When you are departing an airport in Class B airspace most of the work will be done for you. The ground controller will give you a transponder squawk and the tower controller will hand you off to the appropriate terminal controller:

TOWER *"Navion 5327X, contact Bigtown Departure 125.05 now. Have a good day."*

PILOT *"27X going to departure."*

When you are in Class B airspace you must comply with any heading or altitude changes directed by the terminal controller unless you determine that to do so would violate VFR cloud clearance minimums. It is highly unlikely that a controller will assign a specific heading or altitude to a VFR flight, however.

CONTROLLER *"Cessna 1357X, turn right heading 340, traffic your ten o'clock a 737 climbing to 12,000 feet."*

PILOT *"57X has the traffic. Unable 340 due to weather; 270 would keep us clear of clouds."*

CONTROLLER *"Roger, 57X, turn right heading 270 and advise when you can turn north."*

Although you should never turn the safety of your flight over to the controller, what looks like a traffic conflict to you might be part of the controller's master plan.

CONTROLLER *"Cessna 1357X, turn left heading 180, descend and maintain 2,500."*

PILOT *"57X, is the 747 at my two o'clock below 2,500?"*

CONTROLLER *"Affirmative, 57X, the 747 is for runway 17 left, indicating 2,000 and descending."*

PILOT *"57X, left to 180. Leaving 3,000 for 2,500."*

That's a whole lot better than simply following instructions blindly and showing your Christmas tie to the passengers on the 747.

On the reverse side of the terminal area chart you will find VFR flyways that are designed to let you fly through the area with a minimum of radio contact. Look them over, discuss them with your instructor, and use them as much as possible. If you fly the routes shown at the altitudes indicated on the chart you will be safely clear of the big iron. This does not mean that you don't need a clearance to enter the Class B airspace but it does show what you can ask for and almost always get—although it is not a sure thing.

In really congested airspace, such as that around the Los Angeles International Airport, the FAA has designated VFR corridors. Note the difference between a flyway and a corridor: A corridor is not part of the surrounding Class B airspace, and a flyway is. In the case of Los Angeles, the chart shows the altitude, the route of flight, and the frequency to be used to announce the fact that you are in the corridor. No contact with ATC is required or expected. This is not the kind of airspace you want to blunder into unprepared.

Every primary airport with Class B airspace has a control tower, but there is no Class D airspace as such. The terminal controller will hand you off to the tower:

> APPROACH *"57X you're following the twin on right base, contact the Tower 118.3 now."*
>
> PILOT *"57X going to the tower."*

When departing the primary airport in Class B airspace listen closely to the controller and make sure you have a clearance to operate in the Class B airspace. If you're not sure, ask. The assignment of a transponder code and a handoff from the tower to the departure controller do not constitute a clearance if you are operating under Visual Flight Rules.

It helps a lot if you say "VFR" as often as possible:

> PILOT *"Bigtown Departure, Piper 70497, 1,000 climbing 4,500, VFR to Littleburg."*
>
> DEPARTURE *"497, radar contact, cleared to operate in the Bravo airspace — we had you as an instrument departure."*

Departing From a Satellite Airport Beneath Class B Airspace

Look at almost any Terminal Area Chart and you will see airports that are located outside of the Class B surface area, but beneath a "shelf" in the Bravo airspace. Auburn offers a good example (Figure 8-3 and airport J on the sectional chart excerpt). An airplane departing Auburn heading westbound will run right into Class B airspace and must obtain a clearance into that airspace before takeoff or immediately thereafter:

> PILOT *"Seattle Departure, Piper 70497 off from Auburn VFR to Bremerton, request clearance through your Bravo airspace,"*

...and have that clearance in hand before approaching closer than a mile to the Bravo airspace boundary.

Figure 8-3. Auburn Municipal Airport, Washington

Summary

As airspace goes, Class B is more important to the VFR pilot than any other airspace designation. You absolutely, positively, must hear "Cleared to operate in the Bravo airspace" before entry. If you are not sure, ask. "Radar contact" does not constitute a clearance into Class B airspace. You must maintain VFR and fly any headings and altitudes assigned; if you would lose VFR cloud clearance by doing so, you must inform the controller and ask for an amended clearance.

It's for the Chosen Few

You won't be operating in Class A airspace unless you have an instrument rating and are operating on an instrument flight plan, so there are no special communication requirements or procedures for VFR flight in Class A airspace.

Occasionally, a VFR pilot who wins the lottery or gets that big promotion buys an airplane that is capable of flying above 18,000 feet and climbs through that altitude in attempt to get above the weather. Almost every Center controller has a story about an exchange with a pilot who has been legally flying in Class A airspace:

PILOT *"Center, are you talking to that Centurion [or Malibu, or 340, or...] at my two o'clock, about my altitude?"*

CENTER *"Negative. He's squawking VFR and we're not talking to him. Are you close enough to get his tail number?"*

ATC tracks these interlopers to their destinations, of course, and they get a violation.

AUTOMATED FLIGHT SERVICE STATIONS

Service is Their Middle Name

As a VFR pilot you will be talking to flight service stations (FSS) frequently, so they deserve coverage in detail.

The FAA transferred responsibility for the Flight Service function to private industry in 2006 and the jury is still out on how this is working. You will still use 1-800-WX-BRIEF, but **http://www.afss.com/** includes other contact numbers. You will almost certainly be talking to a briefer in a distant city who knows nothing about your local area.

Note: If you use a cell phone when far from home, the system will think you are calling from your phone's home area code. Get a local direct-dial number from the A/FD, or a toll-free number at www.aopa.org/whatsnew/air_traffic/afss_tollfree.html.

To get the same geographical coverage they used to get with multiple manned locations, the FAA uses remote communication outlets (RCOs) and ground communication outlets (GCOs); VORs with voice capability are being phased out.

Air traffic controllers, whether they work in a tower, at a radar facility, or at a flight service station (they all belong to the same union) are at all times alert for incoming calls—that's what they are there for. In most cases the paperwork load at a radar facility is handled by an assistant. Conversely, a flight service station specialist does it all. That is not to say that every specialist does every job—don't ask a Flight Watch specialist to copy your flight plan, and (during Flight Watch hours, *see* Appendix A) don't expect a non-Flight Watch specialist to have the latest weather available.

What Frequency Do I Use?

The Airport/Facility Directory legend pages in the front of each little green book contain a complete explanation of each element of an airport listing. Look for "Communications," and read every word. The legend on the back of a sectional chart is helpful but does not contain the level of detail to be found in the A/FD.

Making Contact

All of those RCOs mean that the specialist has a bank of pushbuttons to monitor, and when you call in, one of those buttons flickers faintly. It doesn't flash on and off as your telephone does when you put someone on hold, it just flickers. An initial transmission that includes your position and the facility you are using for communication is essential. If you transmit,

PILOT *"Riverside Radio, Baron 1014W, over"*

…you present a busy specialist with a puzzle. Which button flickered? If you are at a high enough altitude to reach two remote locations that are assigned the same frequency, both buttons flickered and the specialist needs to know which one to reply on. This is no time for game playing:

PILOT *"Riverside Radio, Baron 1014W on 122.2 over Dagget"*

—is the way to go.

If you want to communicate through an RCO, you'll find its name in a box, with the name of the controlling FSS beneath it (see the box just west of Paine Field, airport F on the sectional chart excerpt). In this situation, you transmit on 122.55:

PILOT *"Seattle Radio, Piper 70497, on 122.55 [give location], extend my flight plan from Sky Harbor to Olympia, new ETA 1830Z."*

This will work even if the specialist is on the phone or otherwise occupied when your call comes in, because you have eliminated all doubt as to what frequency should be used.

A major drawback of the move to automated flight service stations is that in most cases the specialist can no longer look out of the window and provide a pilot with information about traffic in the pattern. At an AFSS, the specialists sit at banks of computer monitors in darkened rooms without windows, just like air traffic controllers at radar facilities.

You can learn of the existence of a GCO only by referring to the A/FD—they are not shown on charts. Where a GCO is installed, you access the AFSS by clicking

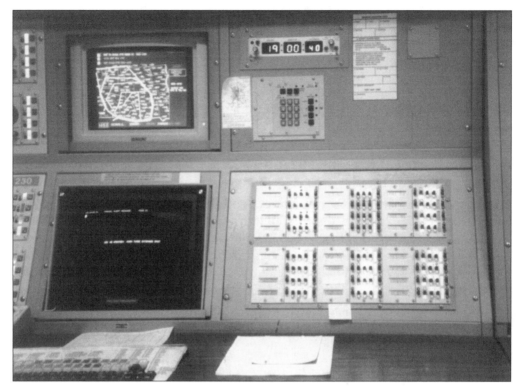

Figure 10-1. Automated Flight Service Station console

the push-to-talk button on your microphone six times; this will establish a telephone connection to the AFSS briefer. Use this facility for last-minute weather briefings and to close flight plans. (Four clicks will connect you to ATC to pick up an instrument clearance or close an IFR flight plan.) GCOs are not to be used to file flight plans.

Special Use Airspace

The AFSS is your best source of information on the status of Special Use Airspace (SUA) such as Restricted, Warning, Alert, and Military Operations Areas. The AFSS specialist also has the latest available information on Military Training Routes (MTRs). I say "latest available" because the military quite frequently changes its plans at the last moment, too late to notify the AFSS.

Be sure to ask about Temporary Flight Restriction areas (TFRs). The FSS is your last line of defense against having an F-16 pull up alongside.

The controlling agency for any given area of SUA is given in the communications panel on sectional charts, and if you want to call the control tower at a military air field to ask if a certain area is "hot" you are free to do so:

> **PILOT** *"Whidbey Tower, Baron 1014W, what is the status of restricted area R-6701?"*

Calling the AFSS will get you exactly the same information:

> **PILOT** *"Seattle Radio, Baron 1014W over Port Townsend on 122.55, what is the status of R-6701?"*

Military aircraft using MTRs are camouflaged and fast, and are therefore hard to see. Planes flying on routes identified with four-digit numbers (IR 1006, VR 1007) will be flying below 1,500 feet above ground level — in some cases as low as 100 feet above the ground. Routes identified with three-digit numbers (IR 008, VR 009) are flown more than 1,500 feet above ground level, but some segments may be below that altitude.

The military advises the AFSS what routes are to be used and the hours during which the route will be hot. Your call to the AFSS will be very much like a check on SUA:

> **PILOT** *"Fort Worth Radio, Baron 1014W, over Childress. What is the status of military route India Romeo 1142?"*

In any event, you should plan to cross military training routes at a 90-degree angle so as to limit your exposure to conflicting military traffic. Military airplanes quite frequently fly in formations that spread out a mile or more from the centerline of the MTR, so paralleling an MTR is a big mistake. Also, if you do catch a glimpse of a military airplane there is probably one or more nearby.

Filing Flight Plans

As soon as you start your student cross-country flights, your instructor begins to drum into your head the need to file VFR flight plans, and to remember to open and close them. The advent of the automated FSS and the proliferation of remote communications sites makes this easy. Dialing 1-800-WX-BRIEF works virtually everywhere.

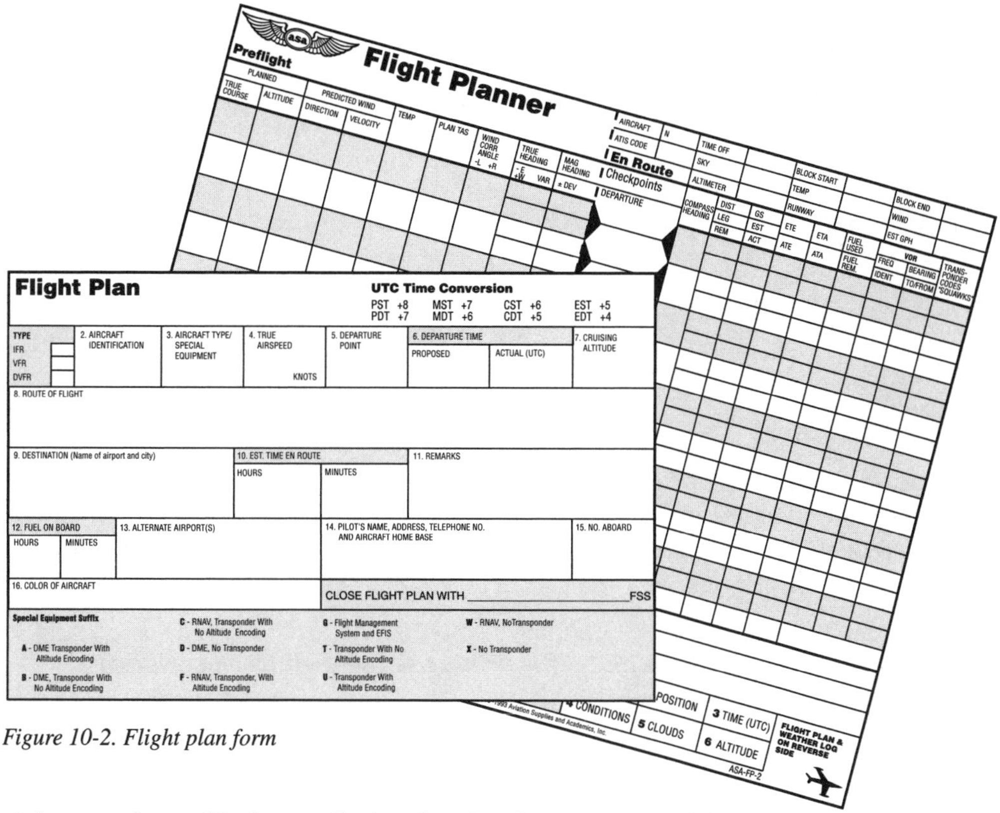

Figure 10-2. Flight plan form

After you have filed your flight plan (with its proposed departure time) with the AFSS specialist, it is held in pending status waiting for you to call and activate it. After making your initial call and establishing communication:

PILOT *"Williamsport Radio, Piper 70497, please open my VFR flight plan from McGinness to Chambersburg at five minutes past the hour."*

If you call more than ten minutes after departure, it is important to give the time that you actually took off, rather than the time at which you call the AFSS to activate your flight plan, because your filed time en route is calculated from the departure airport to the destination airport.

Air-Filing Flight Plans

Flight service stations will accept air-filed flight plans, but they would rather have you do it over the phone to cut down on radio frequency use. You should understand that the specialist has a standard FAA flight plan ready to make the process go as smoothly as possible; it is only necessary for you to transmit the information required to fill in the blocks.

PILOT *"Great Falls Radio, Baron 1014W has a VFR flight plan when you are ready to copy."*

RADIO *"This is Great Falls, go ahead."*

PILOT *"Great Falls Radio, Baron 1014W: VFR, 1014W, Baron E-55 slant Golf, 160 knots, Missoula, off at 0800Z, 12,500 feet, V-2,…"*

You get the idea — you don't have to read the heading of each block to the specialist, just provide the information that goes in the block on the flight plan form.

Position Reports

The practice of making position reports while on a VFR cross-country has fallen into disuse. If you are receiving radar flight following from an ARTCC, of course, position reports are unnecessary — the controller knows where you are. If you are not using that service, however, it is in your best interest to make position reports:

PILOT *"Casper Radio, Cessna 1357X is over Converse County at 15 minutes past the hour, VFR from Casper to Lusk."*

If you fail to arrive at Lusk, the Search-and-Rescue folks can pretty much forget about searching between Casper and Converse County. Search and Rescue is the reason you file flight plans in the first place, so it is in your interest to give potential searchers as much information as possible. Making a pilot report or asking Flight Watch for weather information is a great way to put your position on record.

Many pilots assume that their flight plan is etched in stone and fail to amend it when things don't work out the way they planned. This means that searchers concentrate their efforts along the flight-planned route because they have no way of knowing that the pilot diverted to another route:

PILOT *"Seattle Radio, Piper 70497, over Sunnyside at 4,500 feet. I am on a VFR flight plan from Yakima to Olympia via Snoqualmie Pass. Please amend my route to via the Columbia Gorge to the Battle Ground VOR, then direct Olympia and extend my ETA to 1500Z."*

Close Your Flight Plan!

Except for Emergency Locator Beacon false alarms, nothing causes more headaches for the Search and Rescue folks than failure to close flight plans upon arrival. If you don't contact the FSS and close out your flight plan, the searchers are going to come looking for you—it's that simple. The S&R business operates on a no-news-is-bad-news basis, and they know that every minute counts to an injured pilot awaiting rescue.

If you have the destination airport in sight and are assured of landing, call the AFSS and close your flight plan in the air—but be very sure! Pilots have crashed within a mile or so of the airport after closing their flight plans and no search was begun until relatives began calling for information. If the destination airport has a Remote Communication Outlet, use that or the phone. If you are at a location where 1-800-WX-BRIEF doesn't work, call long distance if you have to, but *close your flight plan!*

Direction Finding Service

Many flight service stations offer direction finding service—check the A/FD and look for VHF-DF (even though the FAA is phasing out DF function from the system and it will soon be gone). When you call one of these stations and report that you are lost or disoriented, the AFSS specialist will ask you to key the push-to-talk switch on your microphone for ten seconds or so—counting from one to ten and back will do the job—and through use of the DF equipment he or she will be able to give you headings to fly toward better weather or an airport. You should use 121.5 MHz (or any AFSS or tower frequency) and say MAYDAY in your initial call if you are in a distress situation:

PILOT *"Mayday, Mayday, Mayday, any station this is Piper 70497. I'm not sure of my position and I'm running low on fuel."*

You will get several answers; respond to the clearest signal.

AFSS specialists must be periodically recertified for direction finding duties and this means they need as much practice as they can get. For this reason, if you are just out tooling around and have 15 minutes or so of free time, call the closest AFSS on 122.2 or the RCO frequency listed on your sectional and say:

PILOT *"Butte Radio, Cessna 1357X, would you like to give me a practice DF steer?"*

They may change you to another frequency so the DF exercise doesn't conflict with other AFSS duties. *Do not* give your location in your initial call; that would take all the challenge out of the problem for the AFSS personnel.

Direction finding is not the only trick up the AFSS specialist's sleeve—they also practice lost aircraft orientation using VOR and ADF.

Can You Tell Me Where There is a Hole?

In this enlightened age there are still VFR pilots who will fly 100 miles to an area which was reporting overcast skies at the time of departure, hoping against hope there will be an opening in the clouds through which they can descend. Almost all of them survive, although they create a lot of adrenaline in the pilots and controllers who monitor the situation on the radio. Sometimes they have to turn around and fly back home, sometimes they find a hole through dumb luck, and sometimes they have to call for help.

The direction finder service mentioned earlier can help, if there is an airport with VFR weather available, but because radar coverage is so extensive, lost pilots usually end up talking to a radar controller who can vector them to a known VFR area (if one exists) or radar-monitor a descent through the clouds to VFR conditions over water or flat terrain.

The military operates Ground Controlled Approach (GCA) facilities throughout the nation, and they are able to talk pilots right down to the ground. Your best bet, if that hoped-for hole doesn't materialize, is to check with the AFSS to learn if there is a GCA available. The military controllers need practice in order to maintain their certification, and solicit the general aviation community for practice approaches. This is something well worth doing with an instructor or safety pilot on board just in case you might need a GCA some day.

Flight Watch

The Enroute Flight Advisory Service, known and addressed as "Flight Watch," has as its sole responsibility the provision of weather information to pilots. A discrete frequency, 122.0 MHz, is assigned to Flight Watch and a specially-trained controller guards that frequency from 0600 until 2200 local time. This frequency is to be used by pilots flying at altitudes up to 17,500 feet MSL; discrete frequencies are available for flights above that altitude (check the A/FD). If you can't make contact, use a regular AFSS frequency.

Flight Watch will provide you with weather information pertinent to your flight—that is, weather for your destination and enroute airports. Requests for random or general weather information will be referred to an appropriate AFSS frequency. Flight Watch is not to be used for filing flight plans or any other non-weather purpose. It is the means through which pilot reports (PIREPs) are collected and disseminated.

The inside back cover of the Airport/Facility Directory for each geographic region contains a map of Flight Watch stations and their remote outlets.

> **PILOT** *"Cedar City Flight Watch, Cessna 1357X over Billings, VFR to Bozeman, can I have the latest Bozeman observation and the terminal forecast, please?"*

— will work every time. If, after reading this book, you still do not own a copy of the A/FD, simply switching to 122.0 and saying,

> **PILOT** *"Flight Watch, Cessna 1357X over Billings, over"*

— will get an answer.

Another ploy works if you know which Air Traffic Control Center's airspace you are flying in: check the communications panel on your sectional for nearby special use airspace. At the end of the listing you will see the appropriate Center frequency.

Pilot Reports

The *Aeronautical Information Manual* contains explicit instructions on how to provide pilot reports to a Flight Watch specialist (between 2200 and 0600 local) or AFSS radio position, and for their benefit you should use the prescribed format. However, they will accept just about anything. If you miss an item they will ask for it. Pilot reports provide critical information to pilots during the flight planning stage, and you should observe the Golden Rule: Do unto others as you would have them do unto you.

If the weather forecast calls for marginal VFR along your route and you are thinking of canceling, wouldn't it be nice to get a PIREP saying that the forecast is wrong and the weather is beautiful? When you are flying in beautiful weather, then, file a pilot report:

> **PILOT** *"Oakland Flight Watch, Piper 70497 over Fresno at 7,500 feet VFR from Delano to Sacramento. Smooth ride all the way, good VFR at this altitude, lower broken layer between Bakersfield and Visalia."*

…or, if the tops of the clouds were forecast to be at 12,000 to 15,000 feet, and you are flying behind a 65 horsepower engine, wouldn't it be nice to learn that they are no higher than 7,000 feet?

It's a two-way street — file pilot reports and ask the AFSS for pilot reports. Your reports will not only be made available to other pilots, but will also be forwarded to the National Weather Service for use in making and correcting forecasts.

Summary

The network of flight service stations is an asset that is not used often enough by general aviation pilots, for reasons that are not clear to me. Filing and closing flight plans is neither difficult nor onerous; as a search pilot I can tell you that the lack of a destination or route on a missing aircraft makes search planning a matter of playing guessing games.

What's the Difference?

In many ways, the procedures and techniques you have learned so far duplicate what you will need to know as you begin training for the instrument rating. The terminology will change a little, but the major difference is that you must be constantly listening for a call from a controller. If you miss a call, the controller must take time away from other pilots to call you again and again. You can't just turn down the volume and enjoy the scenery. Communication is a team effort that involves both the controller and the pilot in achieving the goal of a safe flight. Neither member of the team can afford to sit back and take it easy.

You have many sources for information on frequencies: approach plates, low altitude enroute charts (which include a communications panel), and the A/FD. Most GPS navigators will display the appropriate frequencies automatically when you enter a waypoint or airport designator.

Be sure to check the AIM frequently and stay on top of changes. Changes to the AIM are issued twice a year, and it is your best source of updated information. Be leery of information in old editions of the AIM. With the approval status of GPS and its cousins WAAS and LAAS changing rapidly, you can't afford to rely on what your instructor told you last year.

Filing Your Flight Plan

You have several options when filing: by phone with Flight Service, by DUATS, or by "popping up" en route. Ground communication outlets are not to be used for filing; they provide a means of picking up clearances. Remote communication outlets, on the other hand, can be used for filing.

Unless you are talking directly to Center or a TRACON, keep in mind the 30-minute lead time needed by the Center computer to digest your flight plan and kick out a clearance. The computer must be able to fit your flight into the system, taking into account other flights already cleared over the same route; a radar facility is better able to see the situation in real time and find a slot you can slip into.

When filing, remember that if you lose communication with ATC you are expected to fly the filed route (or as amended by a later clearance). It is better to file a route using airways, and ask for a vector direct when you are airborne, than to file "direct" and have limited options when you are suddenly forced to navigate on your own. GPS is not yet approved as a "sole source" means of navigation (that requires a WAAS-enabled box) — you must have a ground-based backup, and for the foreseeable future that will be VORs.

If you are flying a non-turbine airplane at less than 10,000 feet and the proposed flight will be between metropolitan airports with radar facilities, check your approach plate booklet to see if the airports are listed under Tower Enroute Control — put "TEC" in the Remarks section of the flight plan form. This avoids the delay inherent in going through the Center computer for a routing by keeping you in terminal radar space all the way. You will find that it is possible to use TEC to fly to airports that are not listed but which are close to your route of flight; as long as the terminal facility has you on radar all the way.

OTP

VFR-On-Top (OTP) is a useful tool. If you are departing from a towered airport and know that the tops are fairly low, you can tell ground control that you would like a climb to VFR conditions on top. Ground will coordinate with the facility that has responsibility for the overlying airspace, and if all goes well you should hear,

CONTROLLER *"Baron 1014W is cleared to the Seattle VORTAC. Climb to VFR conditions on top; if not on top at 7,000 feet advise. Squawk 4610."*

The clearance limit will be a navaid or fix close to the departure airport. Of course, you should be on top well before reaching the altitude specified in the clearance. When you break out (and are 1,000 feet above the clouds) you can cancel IFR and proceed VFR.

If the weather is VFR at the departure airport but there is a possibility that you will encounter instrument conditions en route, you can put OTP in the altitude block on the flight plan form. This gets you into the IFR system, but with the freedom to choose your own altitude (MEA or above).

Write It Down

The AIM suggests that pilots keep a written record of all clearances, and forms are available at pilot supply stores for those who don't want to "roll their own." Because route clearances follow a consistent format, let me suggest the old C-R-A-F-T trick. Simply write the letters C, R, A, F, and T vertically along the left-hand edge of a piece of paper; the letters stand for:

C learance limit. Usually (but not always) the destination airport.

R oute. This might be a series of airways and intersections.

A ltitude. This is usually a straight climb but might include restrictions.

F requency. This will be departure control or Center.

T ransponder squawk. Self-explanatory.

If you fly out of one airport regularly, you will be able to fill in most of the blanks even before you start the engine, just based on experience.

C Clearance Limit

If you have filed an IFR flight plan from Dallas to Houston, you have a 90 percent chance of being right if you write HOUSTON, because your clearance will probably begin "ATC clears Baron 1014W to the Houston airport via…" ATC used to clear flights to enroute intersections and have them enter holding patterns at those intersections until traffic permitted them to proceed to their destinations. Now, ATC holds flights on the ground rather than in the air. That's why there is such a small chance of being cleared to an enroue fix.

R Route

If you have filed a preferred route to your destination (*see* the Airport/Facility Directory), you may hear "cleared as filed." If your departure airport has a published departure procedure (DP), you will hear "…via the Lakes One Departure, GONZO, V-15…".

A busy airport may have several instrument departures, with the DP assigned being a function of which way you are going. You could monitor ground or clearance delivery to see what type of departure is being assigned to other flights going your way, or you could simply call the tower on the phone and ask. In any case, you should be able write a route after the R that would require only a little bit of editing as your clearance is transmitted to you.

Ⓐ Altitude

If there is an altitude restriction in the published DP, the controller need not repeat it in the clearance. For example, at Boeing Field the NEEDLE FIVE departure tells the pilot to maintain 2,000 feet and expect higher within 6 miles. In this case I would write 2,000 after the A, followed by an upward slanting arrow with "6mi" written under it, followed by the filed altitude. If, upon receipt, the clearance to a higher altitude is something other than the filed altitude, all I have to do is draw a line through it and write in the new assignment.

If you are departing an airport in uncontrolled airspace, your clearance will include something like "Upon entering controlled airspace (turn right/left, proceed direct to, intercept...)." Obviously, you must listen carefully. Do not accept an "open-ended" clearance. Be sure that you can comply with the clearance in the event of lost communications.

Ⓕ Frequency

You should be able to determine the departure control frequency in use by monitoring the tower for a few minutes. Or look at the A/FD—there probably aren't more than two frequencies; write them both down and draw a line through the one that you are not assigned.

Ⓣ Transponder

If you can predetermine your transponder code you should buy a bunch of lottery tickets. When the Center computer coughs up your clearance it assigns a code. There is no way in the world to anticipate it, but hey! It's only four digits!

Appendix C in this book contains suggested clearance-copying shortcuts.

Ready to Copy

So there you are, with most of your clearance already written down and a pencil ready to correct only those items that differ from your best guess—that's a whole lot better than trying to copy a clearance from scratch on a clean piece of paper:

PILOT *"Oakland Clearance Delivery, Cessna 1357X is ready to copy."*

CLEARANCE DELIVERY *"57X, your clearance is on request."*

What does that mean? Do you have to request it again or something? No, it means that when you taxied out, one of the controllers in the tower picked up the phone and told the Center controller that you would be calling for your clearance soon (requested your clearance). Just as soon as it comes out of the computer, it will be ready for you.

By the way, never fail to mention the fact that you are going IFR to everyone who might be interested:

PILOT *"Oakland Ground, Cessna 1357X at the general aviation ramp, IFR to Seattle."*

PILOT *"Oakland Tower, Cessna 1357X ready for takeoff on 27R at Alfa One, IFR to Seattle."*

"Cleared for Takeoff"

It doesn't matter whether you are taking off from a major airport with a handoff from tower to departure control, from a nontowered airport with a published departure procedure, or from a small strip with a clearance received over the phone, your takeoff procedure will be the same as a VFR departure. The only difference is who you talk to and when you talk to them.

At the major airport, the tower controller has pointed out your data block to the radar controller, who knows who you are and where you are going. You should hear:

TOWER *"Cessna 1357X, contact departure."*

PILOT *"57X, Wilco."*

PILOT *"Seattle Departure, Cessna 1357X, 2,000 climbing 8,000."*

What if you are departing from a smaller airport, one that doesn't have such niceties as terminal radar? The word from here is: *carry sectional charts!* You'll probably be getting your clearance from an RCO, a GCO, or over the phone from the flight service station, and it is going to be something like this:

CONTROLLER *"ATC clears Piper 70497 to the Missoula Airport as filed, maintain runway heading, when able proceed direct Mullan Pass VOR, maintain 9,000 feet, reaching 6,000 feet contact Salt Lake Center 128.35, squawk 6723. Clearance void if not off by 20 minutes past the hour."*

What does "when able" mean? The Pilot/Controller Glossary definition is not very helpful. As a practical matter, it means "when you can proceed without hitting anything..." It doesn't mean "when able to receive Mullan Pass," because it might be possible to get a signal and still be below the tops of the mountains, and it certainly doesn't mean "as soon as you are off the ground." Now you can see why I emphasized the need for a sectional chart; you have to provide your own terrain clearance until you reach 6,000 feet (in this case), and you need to know where the mountains are. Ask ATC, "Are you providing terrain clearance?" if you are unsure.

What does "clearance void after..." mean? The Center computer is reserving a slice of airspace for you, based on your departing before the void time. If you take off just as your watch ticks off the void time, you are too late. If you need extra time to get the airplane warmed up and into position for takeoff, let the FSS briefer know at the time you file.

What if you hear Salt Lake City on 128.35 as you are climbing through 5,000 feet and decide to give them an early call? The controller says "radar contact." Now who is in charge of terrain clearance? You are! Until you have reached the controller's minimum instrument altitude, or MIA (something that the controller knows but you don't), and the controller gives you a vector, the controller has no responsibility to keep you away from the rocks. It's pretty easy to figure that 6,000 feet is the MIA in that controller's sector. The job of an air traffic controller is to keep IFR airplanes separated from one another, period. ATC will not give you a vector until you are so high that there is nothing available for you to hit except other airplanes, and they won't allow that.

If the remote airport you are departing from does not have controlled airspace all the way to the surface, for the first 700 or 1,200 feet of climb you will be in uncontrolled airspace...and controllers can't control in uncontrolled airspace. This is what you may hear as part of your clearance:

> **CONTROLLER** *"Upon entering controlled airspace, cleared as filed."*

Or, after asking which runway you will be departing from (when there is more than one):

> **CONTROLLER** *"Maintain runway heading until entering controlled airspace, then (turn to heading, proceed to navaid/fix, then as filed...)."*

If you should receive such a clearance, with no definite instructions for your action upon entering controlled airspace, ask for clarification; you need to know for "lost comms" purposes.

"Request a Vector To..."

A "vector direct" works like this: You have departed, and are climbing on the first leg of the departure procedure or toward the first fix on your filed route. When you reach the altitude at which you were directed to call Center, you say,

> **PILOT** *"Atlanta Center, Baron 1014W, level 6,000, request vector direct Dallas."*

or you could say,

> **PILOT** *"Atlanta Center, Baron 1014W, level 6,000, request heading until receiving Dallas VOR suitable for navigation."*

This works only in a radar environment, but these days that covers most of the continental United States.

Controllers are becoming more comfortable with GPS, but you probably know more about its use than the controller you are talking to does. The AIM accepts their use for "situational awareness."

On the Way

Once you are airborne, instrument flight is just a matter of flying headings and altitudes assigned. Except for handoffs from one sector to another or from a terminal facility (Approach/Departure) to an enroute facility (Center), you won't have much communicating to do. Understand that a lot goes on behind the scenes; as you approach the boundary between one controller's airspace and the adjacent controller's airspace, the controller you are talking to causes your data block to flash on the scope of the next controller. He or she places a cursor on the flashing data block and—click!—the handoff is accepted. Then, and only then, your controller will tell you to contact the next controller. You do not have to tell the new controller your life story. Be sure to include your altitude in any contact with a new controller:

> **DEPARTURE** *"Baron 1014W, Orlando Departure, contact Jacksonville Center 127.55. Good day."*
>
> **PILOT** *"Going to 127.55, 14W"*

...keep the Departure frequency written down, or on the #2 radio just in case.

> **PILOT** *"Jacksonville Center, Baron 1014W, level 6,000." (or "4,000, climbing 6,000.")*
>
> **CENTER** *"14W, Jacksonville, radar contact, altimeter 30.11."*

When your data block first creeps onto the edge of the controller's scope, using your full callsign is the by-the-book thing to do. After that, when the controller

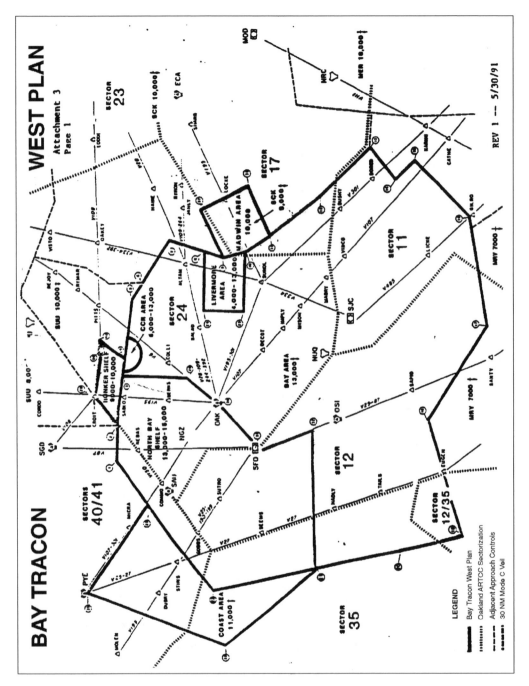

Figure 11-1. This chart is not current, but it illustrates how sectorization works.

is providing IFR separation, combine your callup and your message into one transmission:

> PILOT *"Jacksonville Center, 14W, request lower."*

…instead of calling and waiting for a reply before sending your message.

There are places where the name of the navaid is the same as the name of a nearby airport. When an enroute controller clears you to Podunk, he or she means the Podunk VOR, not the airport.

Weather

Center's radar does not see all of the weather that you see, and you don't see all of the weather that Center sees. Work together. If your course/vector will take you into something nasty, speak up! Quite possibly, it is not being displayed at the controller's position. If your onboard weather display (not radar) shows something bad ahead, ask the controller if he or she sees it. There is at present no "real time" weather in the cockpit except for radar and electrical discharge detectors. Technology isn't perfect yet.

Traffic Reports

When a controller calls traffic, he or she wants to hear either "negative contact," "traffic in sight," or "request vectors around traffic." Saying "we're IMC," "we're in the clouds," or "searching/looking" does not relieve the controller of responsibility for continuing reports until traffic is no longer a factor. Use the correct response, not a weather report.

Sectorization

Both terminal and en route airspace is divided into horizontal and vertical sectors (*see* Figure 11-1). During the quiet hours, staffing levels may call for one controller to take over more than one contiguous sector, while during the morning and evening rush each sector will have a controller assigned. I only mention this to explain why you talk to only one controller at one time of day and to several at another time of day.

Another Handy Trick

For use while en route, make four columns on a 3x5 card headed FREQ, ALT, HDG, and SQUAWK. Every time you are assigned a new frequency, altitude, heading, or squawk, write it in the appropriate column and draw a line through the previous one. That way you can always read the last assignment if you need it and you have the current assignment handy.

Some pilots use the rotatable azimuth on the ADF to remember altitudes—set 40 at the top if your assigned altitude is 4,000 feet, etc.

Holding

Nothing is practiced more during instrument training and used less in real life than holding. When it happens, though, you must be ready. Once they have transmitted holding instructions and they have been acknowledged, controllers do not peer at their scopes to determine what kind of holding pattern entry the pilot has utilized. Chances are that the controller will ignore the flight until the reason for the hold is no longer valid and a further clearance can be issued.

Holding instructions are always issued following a strict format:

- The cardinal direction of the holding course from the fix (north, southwest, etc).
- The name of the fix.
- The radial, course, bearing, airway, etc. on which to hold.
- Leg length in miles for a DME hold.
- Direction of turns if nonstandard, i.e. left turns.
- Expect further clearance time.

CENTER *"Baron 1014W, hold west of the Podunk VORTAC on the 270 radial, expect further clearance at 40 past the hour, time now 10 past the hour."*

The first item is the simplest, yet it gives instrument pilots the most grief. It tells you the location of the holding course; if you are told to hold west of a facility, your first turn after passing the facility will be to a westerly heading ("west" extends from 247.5° to 292.5°…each cardinal heading is 45° wide). The holding airspace could be either north or south of the holding course in this situation…the holding airspace for right turns would be south of the holding course and it would be north of the holding course for a nonstandard pattern (*see* Figure 11-2). You will *always* fly inbound to the fix. That is, if you are cleared to hold southeast on the 147° radial your inbound course will be 327° with a centered needle.

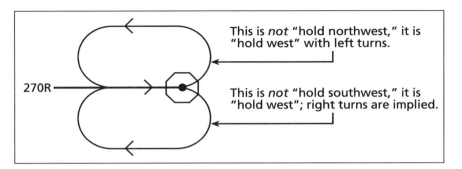

Figure 11-2. "Hold west on the 270° radial"; common misconceptions.

If you are using GPS for navigation, be sure you understand how it handles holds…do you need to suspend sequencing? Do you need to be in LEG or OBS mode? Has the manufacturer made other provisions for this situation?

Expect the Unexpected

Think your instrument flights will always go as planned, smooth as silk? Dream on. What if unexpected headwinds call for an early fuel stop? Airframe ice calls for diversion to more welcoming climes? Just tell your controller.

PILOT *"Fort Worth Center, Baron 1014W requests change destination from Chicago O'Hare to Kansas City Downtown."*

PILOT *"Denver Center, Cessna 1357X is picking up moderate icing; request lower."*

By the way, reporting an icing encounter while flying an airplane not approved for flight in known icing conditions will not get you into any kind of trouble, and it will help other pilots avoid the area. Take immediate action to escape the icing conditions, though. Don't hesitate to declare an emergency (that won't get you into trouble either):

PILOT *"Seattle Center, 1357X picking up ice, request lower."*

CONTROLLER *"57X, unable lower at this time. Expect lower altitude in ten miles."*

PILOT *"Center, 57X is declaring an emergency, leaving 9,000 feet."*

You are covered by your PIC emergency authority. If your action requires that you be given priority over other aircraft, the manager of the ATC facility *may* ask for an explanation.

Changing Altitude

The English language has two troublesome prepositions: to and for. They are troublesome because they sound like two and four. You can eliminate any confusion by simply omitting the prepositions:

> **PILOT** *"Cessna 1357X, leaving 4,000, climbing 6,000."*

> **PILOT** *"Cessna 1357X, leaving 8,000 descending 4,000."*

Pilot's Discretion

Center might assign a new altitude "at pilot's discretion." This is, more often than not, a descent as you near your destination, and it leaves the decision on when to start down in your hands. If there are icy clouds below, it might make sense to stay high until it is absolutely necessary to descend, and then descend as rapidly as possible through the icing layer. This clearance should be included in your first contact with a new controller:

> **PILOT** *"Center, Baron 1014W level 14,000 with discretion to 8,000."*

The new controller might tell you to descend to 8,000 or simply acknowledge your report, keeping the discretionary descent in force.

"Say Heading"

It won't happen very often, but a controller might say,

> **CENTER** *"Cessna 1357X, say heading."*

ATC radar shows your airplane's track over the ground, but it does not know what wind correction you are using to maintain that track. The controller might use this information in deciding when to issue a vector, when to turn you inbound, etc. Do not include heading in a transmission unless requested to do so.

Airspeed

The controller knows your ground speed—it is displayed as part of the data block on the radar screen. On occasion, you might be asked to "say airspeed." Your answer should be indicated airspeed, not true airspeed.

Using Your GPS

"Seattle Departure, Cessna 1357X, request heading 140° until receiving Miami VOR suitable for navigation." Is that legal?? Sure it is, if you have an operable VOR or an IFR-certified GPS on board. The IFR certified box is a slam-dunk, because you filed /G, but if all you have is a handheld, the VOR is your primary navigation receiver (wink, wink).

Or you could have filed the latitude and longitude description of the Miami VOR because the Seattle Center computer probably does not have Miami in its database. It's a sure thing that the Seattle Center computer does not know where Tamiami airport is.

If your destination out of Seattle was Tamiami, you could file KSEA..N2539/W8026..KTMB and note "Lat/long is KTMB" in the Remarks section of the flight plan form.

Cruise Clearances

Are you flying along in the murk, wondering if there's clear air below? Ask ATC for a cruise clearance.

> **PILOT** *"Indianapolis Center, Piper 49470 requests cruise 6,000 feet."*
>
> **CONTROLLER** *"Piper 470, request approved."*

Now you are free to descend from 6,000 feet to the minimum enroute altitude (MEA) for the airway, or the minimum obstruction clearance altitude (MOCA) if one is published, without saying a word to anyone. If flight at the MEA/MOCA means warmer air or visual conditions, simply report that fact to the controller and you will be told to maintain the new altitude. Implicit in a cruise clearance is a clearance to execute any instrument approach at the destination airport—don't expect a separate clearance.

STARs

Standard Terminal Arrival Routes (STARs) are designed for jets and turboprops, but that doesn't mean that you will never be assigned one. Important note: If you are "Cleared via the CIVET FOUR arrival," you can only leave an altitude when cleared, despite what the plate says. If you hear "Descend via the CIVET FOUR arrival," you can follow the altitudes on the plate without further clearance. What a difference a word makes.

Approach Clearances

As you near your destination, the words you are waiting to hear are "cleared for the approach." Until you hear those words you must maintain the last assigned altitude. Once you have your clearance and you are on a segment of a published approach procedure (approach plate), you can begin your descent. On the plate itself, each segment has a course and a minimum altitude assigned, so when you hear the approach clearance you can leave the last assigned altitude and descend to the altitude for the published segment you are on. I don't want to steal your instrument instructor's thunder, but if you fly exactly on the black lines and at the published minimum altitude you are not going to hit anything.

Unlike the brief clearances you have received en route, requiring only that you acknowledge a change in altitude or heading, an approach clearance is a paragraph. This is the clearance for the ILS approach to runway 13R at Boeing Field:

> **APPROACH** *"Baron 1014W turn left heading 100, intercept the localizer and proceed inbound, cleared for the ILS approach to 13 Right, maintain 2,200 until established on the glide slope, contact the tower 120.6 at NOLLA."*

Now don't let me talk you out of writing the whole clearance down and reading it back word for word (although by the time you wrote half of it down the controller would be saying "14W, did you receive your approach clearance?...Baron 1014W, Boeing Tower..." and generally wondering if you had fallen asleep).

I told you earlier that when airborne, only changes in heading and altitude need be read back. That's still good advice, but in this case it should be expanded upon a little.

> **PILOT** *"1014W, left to 100, cleared for the approach, tower at NOLLA."*

What's wrong with that? You were already at 2,200 (I've done this a few hundred times), so there was no change in altitude. You read back the heading change and acknowledged the approach clearance, and the tower frequency is printed on the approach plate.

It is not unusual for a student to ignore the approach plate and repeat the tower frequency incorrectly:

> **PILOT** *"...and contact tower 123.6 at NOLLA, 497"*

> **APPROACH** *"Negative, 497, the tower frequency is 120.6 — 120.6"*

> **PILOT** *"Roger, approach, contact tower on 120.6, 497."*

Just how well do you think 497 was doing at steering 100, maintaining 2,200 feet, and waiting for the ILS needle to center while getting that straightened out? Probably flew right through the localizer. Why not read the frequency directly from the approach plate and be done with it?

My point is that you need not read back every word that is transmitted by the controller. I hate to tell you this, but you will hear professional pilots flying for air carriers and corporations respond to a lengthy clearance with "We'll do all that." That is going too far, in my opinion.

GPS Approaches

You need your ears on when planning a GPS approach; controller terminology changes depending on whether you are receiving radar services or not. A controller might use a combination of radar and non-radar procedures without making this clear to you. This is the basic rule: If you hear "Proceed direct PODNK, cross PODNK at 7000, cleared GPS runway 12 via PODNK transition," you are receiving non-radar handling and might have to perform a course reversal. If you hear "Fly heading 123 degrees, vector to GPS runway 12 final approach course, maintain 7000 until established on final, cleared for the GPS 12," you are receiving radar services and no course reversal is necessary.

Where there is no TAA, make sure that the controller assigns an altitude to maintain. If in doubt, ask...controller GPS training is still in its infancy and has lagged the explosion in the number of GPS navigators in use. You will not be able to rely on the controller to pick up any errors you might make in flying a GPS procedure, so you must know what is expected of you.

There are some communication "tweaks" unique to GPS approaches you should be aware of. We are so used to hearing "Expect vectors to..." or "Turn right heading 240, vector to intercept..." that we might not notice when the controller does *not* say it. When that phrase is omitted, you are flying a non-radar procedure that includes flying to an initial approach fix and possibly executing a procedure turn. You must know how your GPS navigator handles this situation. Without your intervention, it will almost certainly turn inbound at the IAF because "APPR" was armed as soon as you got within 30 miles of the airport and the unit will sequence waypoints automatically. Some manufacturers designate required procedure turns as waypoints in their flight plans and guide you through them, while others require that you suspend sequencing until you reach the inbound leg of the procedure turn. Know what your box will do. Don't wait to be told whether or not to make a procedure turn — you should know — and don't expect hand-holding from ATC when executing a GPS procedure.

You can shortcut some of the "twenty questions" by being proactive; this transmission tells the controller that you have a plan and know how to carry it out:

PILOT *"Piper 70497 requests the GPS 10 approach via TAYLO transition."*

CONTROLLER *"Piper 497 cross TAYLO at 3,000 feet, cleared for the GPS 10 approach."*

Few controllers will reject such a request. Once you have crossed the named fix, descent and navigation are up to you and your magic box.

Another point worth noting when receiving vectors to final: You must be established on the final approach course at least two miles prior to the final approach point or the GPS will suspend sequencing and display "Approach not activated." You have the ability to go to the appropriate page and activate the approach, but it is additional head-down work at a critical time. Let the controller know you must have that two-mile lead when you are being vectored.

Radar Approaches

No instrument approach procedure is more dependent on communication skills than a radar approach, either precision (ground controlled approach) or surveillance. These approaches will be found at military and joint-use airports, and the military controllers welcome the opportunity to hone their skills. Let's go to McChord Air Force Base…no approach plate is required.

PILOT *"McChord Tower, Cessna 1357X, 2,500' over Fox Island, squawking 1200, request practice GCA."*

TOWER *"Roger, 1357X, request approved. Turn right heading 120 to intercept the final approach course runway 16, descend and maintain 2,000 feet, squawk 0400. Contact final controller on 120.1"*

PILOT *"Right to 120, descending 2,000, going to 120.1." "McChord radar, Cessna 1357X, 2,000."*

FINAL CONTROLLER *"Radar contact, turn right heading 145. Do not acknowledge further transmissions. Two miles from final approach course, turn right heading 150, wind 170 at 10 knots, altimeter 3010…One mile from final approach course…Intecepting final approach course, turn right heading 160, begin descent…Five miles from touchdown… Slightly right of course turn left heading 158; slightly above glide path…four miles from touchdown…"*

This continues, with constant minor heading and descent rate corrections, until you reach minimums. You can't let your wheels touch the runway at a military airport, so when the final controller says,

FINAL CONTROLLER *"1357X over the touchdown zone, take over visually"*

…you take off the hood, thank the controller for his services, and go on your way. A surveillance approach differs in that you get no vertical guidance, just headings, distances from the threshold, and suggested altitudes as you get closer…based on 300 feet above runway elevation per mile.

Visual, Contact, and Circling Approaches

Theoretically, the controller does not know what the weather conditions are at your location on the approach; the latest METAR is posted, but most of the time workload does not allow the luxury of looking unless someone asks. Expect to be cleared for the instrument approach in use.

If the wind does not favor the runway served by the approach, you might be cleared to circle:

CONTROLLER *"Baron 1014W, cleared for the ILS 13, circle to land runway 31."*

There are a couple of issues here. Circling minimums are higher than straight-in minimums, and the approach category you use must be based on your actual approach speed; obstacle clearance is only 300 feet when circling; and you might lose sight of the runway while circling. Bad idea at night. Do you have to circle? If you can accept the tailwind, say,

PILOT *"Tower, 1357X will land straight-in runway 13."*

If the weather is bad enough to require an instrument approach, there should not be any conflicting VFR traffic…but, as always, expect the unexpected.

When approaching an airport where you know the lay of the land and can navigate by landmarks, you can beat the minimums by requesting a contact approach. You have to ask…ATC will never offer one. You must have one-mile visibility and remain clear of clouds, but you can descend below published minimums; keep in mind the minimum safe altitude regulations in §91.119. Saying "I was in the process of landing" will not cut it with someone over whose backyard you flew at 200 feet when five miles from the airport.

A visual approach is not an instrument approach, but it does not take you out of the IFR system unless and until you cancel/close your IFR flight plan; don't do it. There is no advantage to canceling IFR early, except that it gets you off of the controller's list of things to do. Be absolutely sure that you are 500 feet beneath the clouds and have three miles visibility before canceling IFR. Reported weather must be basic VFR or better, and you must have the airport or a preceding airplane in sight when you request a visual, which is as simple as:

CONTROLLER *"...cleared for the VOR runway 6 approach, report on initial."*

PILOT *"57X requests a visual."*

There is a subtle trap here...the airplane you are following may also be on a visual, with instructions to follow yet another airplane. When you hear,

CONTROLLER *"Cleared for the visual runway 6, follow the Learjet on final. Do you have that aircraft in sight?"*

Say "No" in most cases. As soon as you acknowledge making eyeball contact with the airplane you are told to follow, responsibility for maintaining separation shifts from the controller to you.

If the lowest cloud layer is at least 500 feet above the controller's minimum vectoring/minimum instrument altitude, he can offer vectors to a visual approach.

Missed Approaches

Controllers experience something called "expectation of landing" syndrome, based on the assumption that you are going to make a successful approach to a full stop. The idea that you might come back to his or her frequency asking for a second approach is abhorrent. When the weather is close to minimums, it doesn't hurt to tell ATC that you might miss the approach...that prepares the controller for the eventuality and might create a gap in the traffic flow into which you can be inserted if your worst fears are realized.

If you reach Decision Altitude or the missed approach point and the world is still hidden from view:

PILOT *"Podunk Tower, 1014W, missed approach"*

—will do the trick. Where terminal radar is available, you will probably be vectored around for another try...the missed approach procedure on the plate will be ignored. If you are being vectored by Center, or flying the approach without radar assistance, the procedure on the plate will come into play in most instances. If ATC can help, they will.

Did you see the approach lights as you added power to go around? Runway markings? Maybe a second try would work. If you saw nothing, however, it is time to go elsewhere rather than burn precious fuel.

> **PILOT** *"Podunk Center, 1014W, missed approach, proceeding to (missed approach fix), change destination to Bigtown Municipal."*

In this situation ATC could be a tower, Center, or Approach Control…whoever you were talking to on final. Bigtown Municipal need not be your filed alternate; you can go wherever you want to go. The controller does not know what you filed as an alternate. Weather and fuel remaining play a large part in this decision. Once you have climbed to the controlling facility's minimum instrument altitude, you will get a clearance to your new destination.

Practice Approaches

The procedures to be followed when requesting practice approaches are contained in AIM 4-3-21, but you should understand that to a controller, the words "practice approach" mean approaches in VFR conditions, wearing a hood. The clearance you will receive in response to such a request will include "Maintain VFR." It really shakes up a controller when he or she points out traffic to a pilot on a practice approach and hears "We're in the clouds."

If you want to shoot a few approaches to stay sharp or stay current, file an IFR flight plan and get an IFR clearance. Include "multiple low approaches" or similar terminology in the Remarks section of the flight plan, and let the approach or center controller know what you want to do upon initial contact.

Lost Communications

When you prepare for the instrument pilot knowledge exam, and possibly during the oral portion of your practical test, you will have to know what §91.185 requires of a pilot on an IFR flight plan who has suffered a loss of two-way radio communications: set your transponder to 7600, fly at the highest of (1) the altitude assigned in the last ATC clearance received, (2) the Minimum Enroute Altitude for the route segment you are on, or (3) the altitude you have been told to expect in a further clearance. You are also expected to follow the route assigned in your last clearance; the direct route from the point of failure to the fix, route, or airway specified in a vector clearance; or your flight planned route. Section 91.185 also establishes a means of determining when to begin the approach at your destination relative to your estimated time of arrival.

The unofficial word from Center and Terminal controllers across the country is this: When you arrive at your destination, perform the instrument approach procedure of your choice without delay. Don't get involved in holding at a fix until a predetermined time has elapsed, just get the airplane on the ground. The ATC folks know you have no communication capability and have sterilized the airspace around the destination airport so it is all yours. Any delay you insert into the situation will simply further delay those flights holding in the air or on the ground, waiting for you to land.

"Minimum Fuel"

When you are on an instrument flight plan, remember that you are required to have the required fuel reserves on board. If excess vectoring is going to eat into your reserves, do not hesitate to declare "minimum fuel" to the controller (AIM Pilot/Controller Glossary). This is not the declaration of an emergency, but a "heads up" to the controller that delays might cause a problem.

Good Operating Practices

There are some communications issues that do not fall readily into categories such as "departure," "en route," or "approach" but apply at all times:

The Hearback Problem

When you receive an instrument clearance on the ground before departure, you are expected to read it back to the controller; when airborne, you are expected to read back those portions of clearances involving altitudes, altitude restrictions, and vectors. We all think of this as a fail-safe method, but it has a fatal flaw—the very human tendency to hear what we expect to hear.

CONTROLLER *"Baron 1014W, turn right heading 330"*

PILOT *"Right to 300, 14W"*

You would expect the controller to pick up that error, wouldn't you? But the controller expected 14W to say 330 and that is what he heard. It works both ways—if a pilot expects a descent to 2,200 feet and is instead cleared to 3,200 feet, it is altogether possible for that pilot to descend through 3,200 feet on the way to 2,200 feet. The hearback problem has been the suspected cause of several accidents and is the proven cause of hundreds of altitude or heading violations.

Keep It Brief, But Not Too Brief

Frequency time is a precious commodity for an air traffic controller; you must be brief without being incomprehensible. If you shorten your transmissions too much, figuring out what you want will squander the controller's supply of minutes; adding irrelevant information to your transmissions does the same thing. It is a balancing act. You are required to read back those portions of clearances containing altitude assignments or vectors—you are not required to parrot every word the controller says. Once ATC has addressed you by a shortened callsign (57X instead of Cessna 1357X), these five words will work 90 percent of the time:

> **PILOT** *"Cessna Three Seven Xray, Wilco."*

The Last Shall Be First

The tower controller told you to maintain runway heading, but the departure controller told you not to fly over the threshold...what's going on? The answer is that the most recent instruction takes precedence over prior instructions; in this case, the departure controller may be aware of an impending traffic conflict of which the tower controller is unaware. If a restriction is to remain in force, it will be repeated by subsequent controllers.

Silence is Not Agreement

When juggling traffic, controllers must keep a lot of balls in the air at the same time. A controller cannot and does not take the time to monitor the actions of each airplane in the sector. That's why you, as a pilot, have to hold up your end of the implicit controller/pilot partnership for safety. If the controller calls Piper 1357X and you are flying Cessna 1357X (no two airplanes have the same tail number, but controllers can make mistakes as to aircraft models), ask—

> **PILOT** *"Was that for Cessna 1357X?"*

—rather than wait for the controller to finally see the light. Similarly, if you are given a vector in an unexpected direction without explanation ("this is a vector through the localizer for spacing"..."Turn right heading 090 for traffic") don't just sit there waiting for the controller to see and correct the error. Speak up. And help other pilots by exercising the Golden Rule:

`PILOT` *"Departure, I think the wrong airplane acknowledged your last transmission."*

`PILOT` *"Center, Baron 1014W read back the wrong altitude."*

Another situation where you should speak up is a vector to intercept a localizer or an airway:

`CONTROLLER` *"Baron 1014W, turn left heading 130 to intercept the localizer."*

`CONTROLLER` *"Cessna 1357X, turn right heading 240 to intercept V-23."*

Should the Baron pilot turn inbound on the localizer? Should the Cessna pilot turn to follow the airway? Not without an approach clearance, in the case of the Baron, and not without further clarification in the case of the Cessna. When in doubt....

`PILOT` *"Approach, 14W is intercepting the localizer,"*

`PILOT` *"Center, do you want 1357X to join the airway?"*

It's much easier to get things straight in the first place than it is to recover from an incorrect assumption.

EXCERPTS FROM CALLBACK NEWSLETTER

"I READ BACK THE CLEARANCE. NO ADVERSE WORD FROM THE CONTROLLER, SO WE AGREED THAT WE HAD BEEN CLEARED..."

"WE TALKED AMONG OURSELVES THAT IT SEEMED TOO EARLY TO BE GIVEN A DESCENT DOWN TO 6,000 FEET. HOWEVER, SINCE THE CONTROLLER HAD NOT CONTRADICTED OUR READBACK, WE DESCENDED..."

"WE HEARD THE CONTROLLER CALLING ABC 143 FOUR TIMES. WE DID NOT ANSWER THESE CALLS. LATER, WE CAME TO BELIEVE THAT SHE HAD BEEN TRYING TO REACH US (ABC123)."

I'll Show 'Em!!

There are very, very few shy pilots. Almost all pilots are assertive to some degree and some are downright nasty. If you are heading 270° and the controller says:

> **CONTROLLER** *"Cessna 1357X, turn right heading 240"*

…you could make a 330° right turn (into following traffic?) just to make a point. It's a lot easier to say,

> **PILOT** *"A right turn is the long way around for 57X."*

Occasionally, of course, the controller will want you to turn the long way around — but that will be stated in the directive:

> **CONTROLLER** *"Cessna 1357X, turn right heading 240, the long way around."*

It doesn't happen often, but if a pilot doesn't emphasize the fact that the flight is IFR, the tower can issue a takeoff clearance and not realize that it is an IFR flight until the pilot says,

> **PILOT** *"Tower, do you want me to change over to Departure Control now?"*

That's because at busy towers the local controller doesn't get involved in the issuing of clearances or taxi instructions.

Summary

I want to emphasize again that safe flight under instrument flight rules is a joint effort between the pilot and the controller. Controllers are busy, handling many flights, and when they issue a clearance or an instruction they cannot take time out to monitor the pilot's actions; in most cases, the scale of their scopes would make it impossible to pick up small deviations if they did have the luxury of time. Controllers properly expect that rated instrument pilots know what to do and how to do it; this means that when you do not understand what you are asked to do you must request clarification…you are still expected to know how to comply.

Teamwork

I hope that this book has encouraged you to pay more attention to the way you communicate as a pilot. The most important message I want you to take away from this text is this: Aviation communications is a team effort. The folks at the flight service stations, the terminal radar facilities, and the enroute traffic control centers want you to complete your flight safely every bit as much as you do.

When All Else Fails

Remember that the microphone is not a flight control, and that for a VFR pilot there is no such thing as loss of communication. Many pilots fly happily without any radios at all. If your radio/mike/speaker/headset fizzles out in a cloud of acrid smoke you have a lot of options. Setting your transponder to 7600 will alert the world at large to the fact that you cannot communicate normally; this will give controllers a "heads up" if your destination is in Class D airspace.

Section 4-2-13 of the *Aeronautical Information Manual* covers the procedures a VFR pilot should use to join the traffic pattern at an airport in Class D airspace. Basically, you are to remain outside of or above the Class D surface area until you have determined the direction of traffic flow, and then join the pattern while watching for light signals. You can see how squawking 7600 ahead of time would help the controllers in this situation.

My preference would be to land at an uncontrolled airport and contact the destination control tower by phone, giving your airplane type and color, your estimated time of arrival, and the direction from which you will be approaching. That lets the controller advise other pilots in the pattern of your position and intentions when you come into view.

4-2-13. Communications with Tower when Aircraft Transmitter or Receiver or Both are Inoperative

a. Arriving Aircraft

1. Receiver inoperative.

(a) If you have reason to believe your receiver is inoperative, remain outside or above the Class D surface area until the direction and flow of traffic has been determined; then, advise the tower of your type aircraft, position, altitude, intention to land, and request that you be controlled with light signals.

REFERENCE—
AIM, Traffic Control Light Signals, Paragraph 4-3-13.

(b) When you are approximately 3 to 5 miles from the airport, advise the tower of your position and join the airport traffic pattern. From this point on, watch the tower for light signals. Thereafter, if a complete pattern is made, transmit your position downwind and/or turning base leg.

2. Transmitter inoperative.
Remain outside or above the Class D surface area until the direction and flow of traffic has been determined; then, join the airport traffic pattern. Monitor the primary local control frequency as depicted on Sectional Charts for landing or traffic information, and look for a light signal which may be addressed to your aircraft. During hours of daylight, acknowledge tower transmissions or light signals by rocking your wings. At night, acknowledge by blinking the landing or navigation lights. To acknowledge tower transmissions during daylight hours, hovering helicopters will turn in the direction of the controlling facility and flash the landing light. While in flight, helicopters should show their acknowledgement of receiving a transmission by making shallow banks in opposite directions. At night, helicopters will acknowledge receipt of transmissions by flashing either the landing or the search light.

3. Transmitter and receiver inoperative.
Remain outside or above the Class D surface area until the direction and flow of traffic has been determined; then, join the airport traffic pattern and maintain visual contact with the tower to receive light signals. Acknowledge light signals as noted above.

b. Departing Aircraft.
If you experience radio failure prior to leaving the parking area, make every effort to have the equipment repaired. If you are unable to have the malfunction repaired, call the tower by telephone and request authorization to depart without two-way radio communications. If tower authorization is granted, you will be given departure information and requested to monitor the tower frequency or watch for light signals as appropriate. During daylight hours, acknowledge tower transmissions or light signals by moving the ailerons or rudder. At night, acknowledge by blinking the landing or navigation lights. If radio malfunction occurs after departing the parking area, watch the tower for light signals or monitor tower frequency.

REFERENCE—
14 CFR Section 91.125 and 14 CFR Section 91.129.

Figure 12-1. AIM paragraph 4-2-13

You could try this if your destination airport was in Class C airspace, although communication is the heart and soul of that class of airspace, and you are not supposed to enter it without talking to ATC. The worst thing that could happen is refusal to let you enter the airspace. Don't even think about it if Class B airspace is involved.

In Conclusion...

Controllers don't play "gotcha!" and they don't expect perfection. If you are confused or puzzled by a clearance or instruction, don't do anything until you have asked for clarification (however, instructions like "Turn right *immediately!*" should be obeyed without hesitation). "What do you want me to do?—I don't understand," is a phrase that you won't find in many texts, but it will sure work when you need it. And don't forget that old standby, "Unable."

You now have the information you need to communicate competently and confidently in the National Airspace System. I wish you blue skies, tailwinds, and a clear frequency when you need it.

Airport Advisory Area   FSS

At (the very few) airports that have a flight service station on the field, FSS personnel will provide traffic advisories to pilots operating within ten miles of the airport; this is the Airport Advisory Area. The Sectional Chart Legend indicates that 123.6 MHz should be used for this purpose. Note that the number of flight service stations where personnel have the time or capability to look out of the window is rapidly dwindling as the move to automation continues. When calling for Airport Advisory Service, use "Radio" as a callsign.

Aeronautical Advisory Service   UNICOM

This is the official name for what all pilots know as UNICOM. The frequency for UNICOM is in the Airport/Facility Directory (A/FD) under the city listing, as well as on sectional charts in the airport data block. UNICOM stations are operated by airport businesses as a service to pilots and have no official status—do not expect official weather reports or takeoff/landing clearances from UNICOM. In fact, don't be surprised if no one answers your call. The UNICOM frequency is, of course, the Common Traffic Advisory Frequency at an airport with a UNICOM, whether anyone is monitoring the frequency or not.

Air Route Traffic Control Center (Center)   CENTER

These enroute radar facilities exist to maintain separation between IFR flights and between IFR flights and known VFR flights. Centers will provide VFR traffic advisories on a workload-permitting basis, and you can request Flight Following from these folks. Center frequencies are found in the A/FD and on instrument enroute charts; in a pinch, you could call the nearest flight service station and ask for the Center frequency in your area.

Approach/Departure Control   `APPROACH` / `DEPARTURE`

These are positions at a terminal radar facility (also called a TRACON), responsible for the handling of IFR flights to and from the primary airport. Where Class B airspace exists, these are the controllers you call for clearance to operate in that airspace and you can also call these folks to request Flight Following. Frequencies are found in the A/FD under the city name, on sectional charts in the communications panel, and on terminal area charts. All terminal radar facilities are "approach" facilities, and in the absence of specific direction ("Contact Departure 120.4"), always address them as "Approach."

Note: There are some combined tower/approach control facilities that do not have radar. You find the appropriate frequency in the same way, and the only difference you will notice is that the approach controller will ask for frequent position reports over known landmarks so you can be separated from other known traffic.

Automatic Terminal Information Service (ATIS)   `ATIS`

This is a continuous broadcast of an audio tape prepared by an ATC controller that contains wind direction and velocity, temperature, altimeter setting, runway and approach in use, and other information of interest to pilots such as runway or taxiway restriction, obstructions in the area, etc. The ATIS frequency is found in the A/FD under the city name, on sectional charts in the airport data block and in the communications panel, and on terminal area charts.

Clearance Delivery   `CLEARANCE DELIVERY`

This is a position in the control tower responsible for transmitting instrument clearances to IFR flights. The clearance delivery frequency is found on instrument approach procedure charts. At those airports where a pilot must copy the instrument clearance before taxiing, the notation Cpt, meaning "pre-taxi clearance," is included on the instrument approach procedure charts. At airports in Class B and C airspace where a transponder code must be assigned to each flight, the ATIS will direct pilots to contact Clearance Delivery (or Ground Control, if there is no Clearance Delivery position at that airport) for a clearance which will include the transponder code and departure instructions. The ATIS broadcast will include the frequency to use.

Common Traffic Advisory Frequency (CTAF)   FSS TOWER UNICOM   C

Every civil airport has a Common Traffic Advisory Frequency, found in the A/FD and on sectional charts in the airport data block. The purpose of the CTAF is to provide a single frequency that pilots in the area can use to either contact the controlling facility or broadcast their position and intention to other pilots. The CTAF is found in the A/FD and on sectional charts in the airport data block followed by a white C on a blue or magenta background. At airports without a tower, a flight service station, or UNICOM, the CTAF is 122.9, the MULTICOM frequency. At virtually all tower-controlled airports, the tower frequency becomes the CTAF when the tower is closed.

Enroute Flight Advisory Service (Flight Watch)   FLIGHT WATCH

The Enroute Flight Advisory Service (EFAS) is a position at a flight service station with responsibility for handling weather inquiries from pilots in flight. It operates from 0600-2200 local time and uses 122.00 MHz for flights operating up to 17,500 feet MSL. Discrete frequencies are assigned for flights operating between 18,000 feet and 45,000 feet MSL; these frequencies (and a map of EFAS outlets) can be found on the inside back cover of the A/FD.

(Automated) Flight Service Station   FSS

Flight service stations exist to provide information and services to pilots. The FAA is consolidating the FSS system into a relatively small number of automated flight service stations (*see* Chapter 10). FSS frequencies are found on sectional charts and in the A/FD both under the city name and in a separate listing of AFSS frequencies. On sectionals, frequencies will be found listed above VOR frequency boxes and, when remoted, in separate blue boxes. Read the Sectional Chart Legend carefully—some frequencies are not listed, but their existence is indicated by the thickness of the lines forming the box. Always address a flight service station as "Radio."

Ground Control   GROUND

At tower-controlled airports, Ground Control is a position in the tower responsible for controlling aircraft taxiing to and from the runways. Where parallel runways exist, the Ground Controller has no responsibility for the area between the runways—that belongs to the tower, or local, controller. Ground Control frequencies are found in the A/FD under the city name.

Hazardous In-flight Weather Advisory Service (HIWAS) `HIWAS` Ⓗ

HIWAS is a continuous broadcast of forecast hazardous weather conditions on selected navigational aids. HIWAS has no communication capability; that is, a flight service station cannot break into the HIWAS broadcast to transmit to pilots. VORs that broadcast HIWAS information can be identified on sectional charts by a blue circle with a reversed-out "H" in the upper right-hand corner of the VOR frequency box. In the A/FD, the notation HIWAS will appear in the airport listing under "Radio Aids to Navigation."

MULTICOM `MULTICOM`

The MULTICOM frequency is 122.9 MHz and is intended for use by pilots operating at airports with no radio facilities. The self-announce procedures in the *Aeronautical Information Manual* are especially valuable at these airports. In the A/FD, airports where MULTICOM should be used show 122.9 as the CTAF. On sectional charts, the frequency 122.9 is followed by a white C on a dark background, indicating that it is the CTAF.

Tower `TOWER`

The controller responsible for operations on the runways and in the Class B, C, or D airspace surrounding the airport is officially the "local" controller, although "tower" is the commonly used designation. The tower frequency is found on sectional and terminal area charts in the airport data block and the communications panel, and also in the Airport/Facility Directory under the city listing. When the tower is not in operation, the CTAF is still the tower frequency. The tower controller "owns" the surface of the active runway or runways and any taxiways between active runways, as well as the airspace up to (typically) 2,500 feet AGL and out to the boundaries shown on the chart.

Appendix B
Airspace Definitions

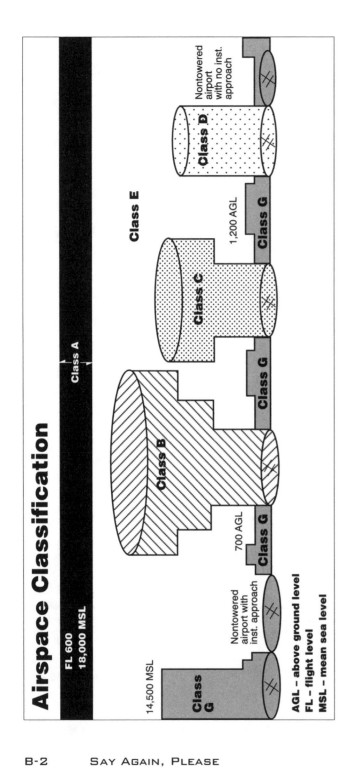

Airspace Classification

Airspace	Class A	Class B	Class C	Class D	Class E	Class G
Entry Requirements	ATC clearance	ATC clearance	Prior two-way communications	Prior two-way communications	Prior two-way communications* or nothing	Prior two-way communications* or nothing
Minimum Pilot Qualifications	Instrument Rating	Private or Student certification. Location dependent.	Student certificate	Student certificate	Student certificate	Student certificate
Two-way Radio Communications	Yes	Yes	Yes	Yes	Yes*	Yes*
Special VFR allowed	No	Yes	Yes	Yes	Yes	N/A
VFR Visibility Minimum	N/A	3 statute miles	3 statute miles	3 statute miles	3 statute miles	1 statute mile
VFR Minimum Distance from Clouds	N/A	Clear of clouds	500' below, 1,000' above, 2,000' horizontal	500' below, 1,000' above, 2,000' horizontal	500' below, 1,000' above, 2,000' horizontal	Clear of clouds
VFR Aircraft Separation	N/A	All	IFR	Runway Operations	None	None
Traffic Advisories	Yes	Yes	Yes	Workload permitting	Workload permitting	Workload permitting
Airport Application	N/A	• Radar • Instrument Approaches • Weather • Control Tower • High Density	• Radar • Instrument Approaches • Weather • Control Tower	• Instrument Approaches • Weather • Control Tower	• Instrument Approaches • Weather	• Control Tower or nothing

*Only if a temporary tower, or control tower is present, which is the exception.

The following shorthand system is recommended by the Federal Aviation Administration (FAA). Applicants for the Instrument Rating may use any shorthand system, in any language, which ensures accurate compliance with air traffic control (ATC) instructions. No shorthand system is required by regulation and no knowledge of shorthand is required for the FAA Knowledge Test; however, because of the vital need for reliable communication between the pilot and controller, clearance information should be unmistakably clear.

The following symbols and contractions represent words and phrases frequently used in clearances. Most of them are used regularly by ATC personnel. By practicing this shorthand, omitting the parenthetical words, you will be able to copy long clearances as fast as they are read.

Example: CAF ➡ RH RV V18 ↑40 SQ 0700 DPC 120.4 Cleared as filed, maintain runway heading for radar vector to Victor 18, climb to 4,000, squawk 0700, departure control frequency is 120.4.

Words and Phrases	Shorthand
Above	ABV
Above (altitude, hundreds of feet)	<u>70</u>
Adjust speed to 250 knots	250 K
Advise	ADZ
After (passing)	<
Airway (designation)	V26
Airport	A
Alternate instructions	( )

Altitude 6,000-17,000	60-170
And	&
Approach	AP
Approach Control	APC
Area Navigation	RNAV
Arriving	↓
At	@
At or above	$\perp$
At or below	$\top$
(ATC) advises	CA
(ATC) clears or cleared	C
(ATC) requests	CR
Back course	BC
Bearing	BR
Before (reaching, passing)	>
Below	BLO
Below (altitude, hundreds of feet)	$\overline{70}$
Center	CTR
Clearance void if not off by (time)	v<
Cleared as filed	CAF
Cleared to airport	A
Cleared to climb/descend at pilot's discretion	PD
Cleared to cross	X
Cleared to depart from the fix	D
Cleared to the fix	F
Cleared to hold and instructions issued	H
Cleared to land	L
Cleared to the outer marker	O
Climb to (altitude, hundreds of feet)	↑70
Contact Approach	CT
Contact (Denver) Approach Control	(den
Contact (Denver) Center	(DEN

Hold (direction) .... H-W

Holding pattern .... ⬭

ILS Approach .... ILS

Increase speed 30 knots .... +30 K

Initial Approach I

Instrument departure procedure .... DP

Intersection .... XN

Join or intercept airway/Jet route/Track or course ..... ⋝

Left turn after takeoff .... ⌡

Locator outer marker .... LOM

Magnetic .... M

Maintain .... M→

Maintain VFR conditions on top .... VFR

Middle compass locator .... ML

Middle marker .... MM

Missed approach .... MA

Nondirectional beacon approach .... NDB

Out of (leave) Control Area .... △

Outer marker .... OM

Over (station) .... OKC

On course .... OC

Precision approach radar .... PAR

Procedure turn .... PT

Radar vector .... RV

Radial (080° radial) 080R

Reduce speed 20 knots .... -20 K

Remain this frequency .... RTF

Remain well to left side .... LS

Remain well to right side .... RS

Report crossing .... RX

Report departing .... RD

Report leaving .... RL

Report on course ....R-CRS

Report over .... RO

Report passing.. RP

Report reaching...RR

Report starting procedure turn.......................................RSPT

Reverse course..RC

Right turn after takeoff...⌐

Runway heading ....RH

Runway (number)...RY18

Squawk..SQ

Standby..STBY

Straight-in approach..SI

Surveillance radar approach .... ASR

Takeoff (direction)..T→N

Tower ....Z

Turn left..TL

Turn right ....TR

Until ..../

Until advised (by) .... UA

Until further advised... UFA

VFR conditions on top... OTP

Via .... VIA

Victor (airway number)... V14

Visual approach.. VA

VOR..⊙

VOR approach..VR

VORTAC..Ⓣ

While in control area ....△

abeam. An aircraft is *abeam* a fix, point, or object when that fix, point, or object is approximately 90 degrees to the right or left of the aircraft track. Indicates a general position rather than a precise point.

abort. To terminate a preplanned aircraft maneuver; e.g., an aborted takeoff.

acknowledge. Let me know that you have received my message.

advise intentions. Tell me what you plan to do.

Aeronautical Information Manual **(AIM).** A primary FAA publication with the purpose of instructing airmen about operating in the U.S. National Airspace System. It provides basic flight information, ATC procedures and general instructional information concerning health, medical facts, factors affecting flight safety, accident and hazard reporting, and types of aeronautical charts and their use.

A/FD. *See* Airport/Facility Directory.

affirmative. Yes.

AFSS. *See* automated flight service station.

Air Force One (or AF Two). Aircraft identification for the United States President or Vice President.

AIRMET. In-flight weather advisories issued only to amend the area forecast, concerning weather phenomena of operational interest to all aircraft and potentially hazardous to aircraft having limited capability because of lack of equipment, instrumentation, or pilot qualifications. AIRMETs are concerned with weather of less severity than that covered by SIGMETs or Convective SIGMETs. AIRMETs cover moderate icing, moderate turbulence, sustained winds of 30 knots or more at the surface, widespread areas of ceilings less than 1,000 feet and/or visibility less than 3 miles, and extensive mountain obscurement.

airport advisory area. The area within ten miles of an airport without a control tower or where the tower is not in operation, and on which a flight service station is located. "[AFSS name] Radio." Accessed via 123.6 MHz.

Airport/Facility Directory (A/FD). A publication designed primarily as a pilot's operational manual containing all airports, seaplane bases, and heliports open to the public including communications data, navigational facilities, and certain special notices and procedures. This publication is issued in seven volumes according to geographical area.

airport surveillance radar (ASR). Approach control radar used to detect and display an aircraft's position in the terminal area. ASR provides range and azimuth information but does not provide elevation data. Coverage of the ASR can extend up to 60 miles.

air route surveillance radar. Air route traffic control center (ARTCC) radar used primarily to detect and display an aircraft's position while en route between terminal areas. The ARSR enables controllers to provide radar air traffic control service when aircraft are within the ARSR coverage. In some instances, ARSR may enable an ARTCC to provide terminal radar services similar to but usually more limited than those provided by a radar approach control.

Air Route Traffic Control Center (ARTCC). A facility established to provide air traffic control service to aircraft operating on IFR flight plans within controlled airspace and principally during the enroute phase of flight. When equipment capabilities and controller workload permit, certain advisory/assistance services may be provided to VFR aircraft. "Center."

air traffic control (ATC). A service operated by appropriate authority to promote the safe, orderly and expeditious flow of air traffic.

approach clearance. Authorization by ATC for a pilot to conduct an instrument approach. The type of instrument approach for which a clearance and other pertinent information is provided in the approach clearance when required.

approach control. Position at a terminal radar facility responsible for handling IFR flights to the primary airport.

ARINC. An acronym for Aeronautical Radio, Inc., a corporation largely owned by a group of airlines. ARINC is licensed by the FCC as an aeronautical station and contracted by the FAA to provide communications support for air traffic control and meteorological services in portions of international airspace.

ARSR. *See* air route surveillance radar.

ARTCC. *See* Air Route Traffic Control Center.

ASR. *See* airport surveillance radar.

ATC. *See* air traffic control.

ATC advises. Used to prefix a message of non-control information when it is relayed to an aircraft by other than an air traffic controller.

ATC clears. Used to prefix an ATC clearance when it is relayed to an aircraft by other than an air traffic controller.

ATC requests. Used to prefix an ATC request when it is relayed to an aircraft by other than an air traffic controller.

ATIS. *See* Automatic Terminal Information Service.

automated flight service station (AFSS). Provides information and services to pilots, using remote communications outlets (RCO) and ground communications outlets (GCO). "[facility name] Radio."

automated UNICOM. Provides completely automated weather, radio check capability and airport advisory information on an automated UNICOM system. These systems offer a variety of features, typically selectable by microphone clicks, on the UNICOM frequency. Availability will be published in the Airport/Facility Directory and approach charts.

Automatic Terminal Information Service (ATIS). The continuous broadcast of recorded non-control information in selected terminal areas. Its purpose is to improve controller effectiveness and to relieve frequency congestion by automating the repetitive transmission of essential but routine information; e.g., "Los Angeles information Alfa. One three zero zero Coordinated Universal Time. Weather measured ceiling two thousand overcast, visibility three, haze, smoke, temperature seven one, dew point five seven, wind two five zero at five, altimeter two niner niner six. I-L-S Runway Two Five Left approach in use, Runway Two Five Right closed, advise you have Alfa."

back-taxi. A term used by air traffic controllers to taxi an aircraft on the runway opposite to the traffic flow. The aircraft may be instructed to back-taxi to the beginning of the runway or at some point before reaching the runway end for the purpose of departure or to exit the runway.

blind transmission. *See* transmitting in the blind.

blocked. Phraseology used to indicate that a radio transmission has been distorted or interrupted due to multiple simultaneous radio transmissions.

broadcast. Transmission of information for which an acknowledgement is not expected.

call up. Initial voice contact between a facility and an aircraft, using the identification of the unit being called and the unit initiating the call.

Center. *See* Air Route Traffic Control Center.

circle to runway [runway number]. Used by ATC to inform the pilot that he must circle to land because the runway in use is other than the runway aligned with the instrument approach procedure. When the direction of the circling maneuver in relation to the airport/runway is required, the controller will state the direction (eight cardinal compass points) and specify a left or right downwind or base leg as appropriate; e.g., "Cleared VOR Runway Three Six Approach circle to Runway Two Two," or "Circle northwest of the airport for a right downwind to Runway Two Two."

Class A airspace. Generally, that airspace from 18,000 feet MSL up to and including FL600, including the airspace overlying the waters within 12 nautical miles of the coast of the 48 contiguous states and Alaska. Unless otherwise authorized, all persons must operate their aircraft under IFR.

Class B airspace. Generally, that airspace from the surface to 10,000 feet MSL surrounding the nation's busiest airports in terms of airport operations or passenger enplanements. The configuration of each Class B airspace area is individually tailored and consists of a surface area and two or more layers (some Class B airspace areas resemble upside-down wedding cakes), and is designed to contain all published instrument procedures once an aircraft enters the airspace. An ATC clearance is required for all aircraft to operate in the area, and all aircraft that are so cleared receive separation services within the airspace. The cloud clearance requirement for VFR operations is "clear of clouds."

Class C airspace. Generally, that airspace from the surface to 4,000 feet above the airport elevation (charted in MSL) surrounding those airports that have an operational control tower, are serviced by a radar approach control, and that have a certain number of IFR operations or passenger enplanements. Although the configuration of each Class C area is individually tailored, the airspace usually consists of a surface area with a 5 nautical mile (NM) radius, an outer circle with a 10 NM radius that extends from 1,200 feet to 4,000 feet above the airport elevation and an outer area. Each person must establish two-way radio communications with the ATC facility providing air traffic services prior to entering the airspace and thereafter maintain those communications while within the airspace. VFR aircraft are only separated from IFR aircraft within the airspace.

Class D airspace. Generally, that airspace from the surface to 2,500 feet above the airport elevation (charted in MSL) surrounding those airports that have an operational control tower. The configuration of each Class D airspace area is individually tailored and when instrument procedures are published, the airspace will normally be designed to contain the procedures. Arrival extensions for instrument approach procedures may be Class D or Class E airspace. Unless otherwise authorized, each person must establish two-way radio communications with the ATC facility providing air traffic services prior to entering the airspace and thereafter maintain those communications while in the airspace. No separation services are provided to VFR aircraft.

Class E airspace. Generally, if the airspace is not Class A, Class B, Class C, or Class D, and it is controlled airspace, it is Class E airspace. Class E airspace extends upward from either the surface or a designated altitude to the overlying or adjacent controlled airspace. When designated as a surface area, the airspace will be configured to contain all instrument procedures. Also in this class are Federal airways, airspace beginning at either 700 or 1,200 feet AGL used to transition to/from the terminal or enroute environment, enroute domestic, and offshore airspace areas designated below 18,000 feet MSL. Unless designated at a lower altitude, Class E airspace begins at 14,500 MSL over the United States, including that airspace overlying the waters within 12 nautical miles of the coast of the 48 contiguous states and Alaska, up to, but not including 18,000 feet MSL, and the airspace above FL600.

Class G airspace. That airspace not designated as Class A, B, C, D or E.

clear of the runway. A taxiing aircraft, which is approaching a runway, is clear of the runway when all parts of the aircraft are held short of the applicable holding position marking. A pilot or controller may consider an aircraft, which is exiting or crossing a runway, to be clear of the runway when all parts of the aircraft are beyond the runway edge and there is no ATC restriction to its continued movement beyond the applicable holding position marking. Pilots and controllers shall exercise good judgement to ensure that adequate separation exists between all aircraft on runways and taxiways at airports with inadequate runway edge lines or holding position markings.

clearance delivery. Control tower position responsible for transmitting departure clearances to IFR flights, and some VFR flights (at busier airports).

clearance limit. The fix, point, or location to which an aircraft is cleared when issued an air traffic clearance.

clearance void if not off by [time]. Used by ATC to advise an aircraft that the departure clearance is automatically canceled if takeoff is not made prior to a specified time. The pilot must obtain a new clearance or cancel his IFR flight plan if not off by the specified time.

cleared as filed. Means the aircraft is cleared to proceed in accordance with the route of flight filed in the flight plan. This clearance does not include the altitude, DP, or DP transition.

cleared [type of] approach. ATC authorization for an aircraft to execute a specific instrument approach procedure to an airport; e.g., "Cleared ILS Runway Three Six Approach."

cleared approach. ATC authorization for an aircraft to execute any standard or special instrument approach procedure for that airport. Normally, an aircraft will be cleared for a specific instrument approach procedure.

cleared for takeoff. ATC authorization for an aircraft to depart. It is predicated on known traffic and known physical airport conditions.

cleared for the option. ATC authorization for an aircraft to make a touch-and-go, low approach, missed approach, stop and go, or full stop landing at the discretion of the pilot. It is normally used in training so that an instructor can evaluate a student's performance under changing situations.

cleared through. ATC authorization for an aircraft to make intermediate stops at specified airports without refiling a flight plan while en route to the clearance limit.

cleared to land. ATC authorization for an aircraft to land. It is predicated on known traffic and known physical airport conditions.

climb to VFR. ATC authorization for an aircraft to climb to VFR conditions within Class B, C, D, and E surface areas when the only weather limitation is restricted visibility. The aircraft must remain clear of clouds while climbing to VFR.

Common Traffic Advisory Frequency (CTAF). A frequency designed for the purpose of carrying out airport advisory practices while operating to or from an airport without an operating control tower. The CTAF may be a UNICOM, MULTICOM, FSS, or tower frequency and is identified in appropriate aeronautical publications.

composite flight plan. A flight plan which specifies VFR operation for one portion of flight and IFR for another portion. It is used primarily in military operations.

compulsory reporting points. Reporting points that must be reported to ATC. They are designated on aeronautical charts by solid triangles, or filed in a flight plan as fixes selected to define direct routes. These points are geographical locations defined by navigation aids/fixes. Pilots should discontinue position reporting over compulsory reporting points when informed by ATC that their aircraft is in "radar contact."

conflict alert. A function of certain air traffic control automated systems designed to alert radar controllers to existing or pending situations between tracked targets (known IFR or VFR aircraft) that require their immediate attention/action.

conflict resolution. The resolution of potential conflictions between aircraft that are radar identified and in communication with ATC by ensuring that radar targets do not touch. Pertinent traffic advisories shall be issued when this procedure is applied.

contact approach. An approach wherein an aircraft on an IFR flight plan, having an air traffic control authorization, operating clear of clouds with at least 1 mile flight visibility and a reasonable expectation of continuing to the destination airport in those conditions, may deviate from the instrument approach procedure and proceed to the destination airport by visual reference to the surface. This approach will only be authorized when requested by the pilot and the reported ground visibility at the destination airport is at least 1 statute mile.

continue. When used as a control instruction should be followed by another word or words clarifying what is expected of the pilot. Example: "continue taxi," "continue descent," "continue inbound," etc.

controlled airspace. An airspace of defined dimensions within which air traffic control service is provided to IFR flights and to VFR flights in accordance with the airspace classification. Controlled airspace is a generic term that covers Class A, Class B, Class C, Class D, and Class E airspace. Controlled airspace is also that airspace within which all aircraft operators are subject to certain pilot qualifications, operating rules, and equipment requirements in Part 91. For IFR operations in any class of controlled airspace, a pilot must file an IFR flight plan and receive an appropriate ATC clearance. Each Class B, Class C, and Class D airspace area designated for an airport contains at least one primary airport around which the airspace is designated.

control sector. An airspace area of defined horizontal and vertical dimensions for which a controller or group of controllers has ATC responsibility, normally within an air route traffic control center or an approach control facility. Sectors are established based on predominant traffic flows, altitude strata, and controller workload. Pilot communications during operations within a sector are normally maintained on discrete frequencies assigned to the sector.

convective SIGMET. A weather advisory concerning convective weather significant to the safety of all aircraft. Convective SIGMETs are issued for tornadoes, lines of thunderstorms, embedded thunderstorms of any intensity level, areas of thunderstorms greater than or equal to VIP level 4 with an area coverage of 4/10 (40%) or more, and hail 3/4-inch or greater.

cross *[fix]* **at** *[altitude]*. Used by ATC when a specific altitude restriction at a specified fix is required.

cross *[fix]* **at or above** *[altitude]*. Used by ATC when an altitude restriction at a specified fix is required. It does not prohibit the aircraft from crossing the fix at a higher altitude than specified; however, the higher altitude may not be one that will violate a succeeding altitude restriction or altitude assignment.

cross *[fix]* **at or below** *[altitude]*. Used by ATC when a maximum crossing altitude at a specific fix is required. It does not prohibit the aircraft from crossing the fix at a lower altitude; however, it must be at or above the minimum IFR altitude.

cruise. Used in an ATC clearance to authorize a pilot to conduct flight at any altitude from the minimum IFR altitude up to and including the altitude specified in the clearance. The pilot may level off at any intermediate altitude within this block of airspace. Climb/descent within the block is to be made at the discretion of the pilot. However, once the pilot starts descent and verbally reports leaving an altitude in the block, he may not return to that altitude without additional ATC clearance. Further, it is approval for the pilot to proceed to and make an approach at the destination airport, and can be used in conjunction with:

   a. An airport clearance limit at locations with a standard/special instrument approach procedure. The regulations require that if an instrument letdown to an airport is necessary, the pilot shall make the letdown in accordance with a standard/special instrument approach procedure for that airport.

   b. An airport clearance limit at locations that are within/below/outside controlled airspace and without a standard/special instrument approach procedure. Such a clearance is *not authorization* for the pilot to descend under IFR conditions below the applicable minimum IFR altitude, nor does it imply that ATC is exercising control over aircraft in Class G airspace. However, it provides a means for the aircraft to proceed to destination airport, descend, and land in accordance with applicable regulations governing VFR flight operations. Also, this provides search and rescue protection until such time as the IFR flight plan is closed.

cruising altitude. An altitude or flight level maintained during enroute level flight. This is a constant altitude and should not be confused with a cruise clearance.

CTAF. *See* Common Traffic Advisory Frequency.

delay indefinite *[reason if known]* **expect further clearance** *[time]*. Used by ATC to inform a pilot when an accurate estimate of the delay time and the reason for the delay cannot immediately be determined; e.g., a disabled aircraft on the runway, terminal or center area saturation, weather below landing minimums, etc.

departure center. The ARTCC having jurisdiction for the airspace that generates a flight to the impacted airport.

departure control. A function of an approach control facility providing air traffic control service for departing IFR and, under certain conditions, VFR aircraft.

DF guidance. Headings provided to aircraft by facilities equipped with direction finding equipment. These headings, if followed, will lead the aircraft to a predetermined point such as the DF station or an airport. DF guidance is given to aircraft in distress or to other aircraft which request the service. Practice DF guidance is provided when workload permits.

DF steer. *See* DF guidance.

direct. Straight-line flight between two navigational aids, fixes, points, or any combination thereof. When used by pilots in describing off-airway routes, points defining direct route segments become compulsory reporting points unless the aircraft is under radar contact.

discrete code. As used in the Air Traffic Control Radar Beacon System (ATCRBS), any one of the 4096 selectable Mode 3/A aircraft transponder codes except those ending in zero zero; e.g., discrete codes: 0010, 1201, 2317, 7777; non-discrete codes: 0100, 1200, 7700. Non-discrete codes are normally reserved for radar facilities that are not equipped with discrete decoding capability and for other purposes such as emergencies (7700), VFR aircraft (1200), etc.

discrete frequency. A separate radio frequency for use in direct pilot-controller communications in air traffic control that reduces frequency congestion by controlling the number of aircraft operating on a particular frequency at one time. Discrete frequencies are normally designated for each control sector in enroute/terminal ATC facilities. Discrete frequencies are listed in the Airport/Facility Directory and the DOD FLIP *IFR Enroute Supplement.*

distress. A condition of being threatened by serious and/or imminent danger and of requiring immediate assistance.

DP. *See* instrument departure procedure.

EDCT. *See* expected departure clearance time.

EFAS. *See* Enroute Flight Advisory Service.

EFC. *See* expect further clearance *[time].*

emergency. A distress or an urgency condition. The emergency VHF frequency is 121.5 MHz.

Enroute Flight Advisory Service (EFAS). A service specifically designed to provide, upon pilot request, timely weather information pertinent to his type of flight, intended route of flight, and altitude. The FSS's providing this service are listed in the Airport/Facility Directory. "Flight Watch." Accessed via 122.0 MHz.

estimated time of arrival (ETA). The time the flight is estimated to arrive at the gate (scheduled operators) or the actual runway on times for nonscheduled operators.

estimated time enroute (ETE). The estimated flying time from departure point to destination (lift-off to touchdown).

ETA. *See* estimated time of arrival.

ETE. *See* estimated time enroute.

execute missed approach. Instructions issued to a pilot making an instrument approach which means continue inbound to the missed approach point and execute the missed approach procedure as described on the Instrument Approach Procedure Chart or as previously assigned by ATC. The pilot may climb immediately to the altitude specified in the missed approach procedure upon making a missed approach. No turns should be initiated prior to reaching the missed approach point. When conducting an ASR or PAR approach, execute the assigned missed approach procedure immediately upon receiving instructions to "execute missed approach."

expect *[altitude]* at *[time]* or *[fix]*. Used under certain conditions to provide a pilot with an altitude to be used in the event of two-way communications failure. It also provides altitude information to assist the pilot in planning.

expected departure clearance time. The runway release time assigned to an aircraft in a controlled departure time program and shown on the flight progress strip as an EDCT.

expect further clearance *[time]* (EFC). The time a pilot can expect to receive clearance beyond a clearance limit.

expect further clearance via *[airways, routes, or fixes]*. Used to inform a pilot of the routing he can expect if any part of the route beyond a short-range clearance limit differs from that filed.

expedite. Used by ATC when prompt compliance is required to avoid the development of an imminent situation. Expedite climb/descent normally indicates to a pilot that the approximate best rate of climb/descent should be used without requiring an exceptional change in aircraft handling characteristics.

FAF. *See* final approach fix.

filed. Normally used in conjunction with flight plans, meaning a flight plan has been submitted to ATC.

final approach fix (FAF). The fix from which the final approach (IFR) to an airport is executed and which identifies the beginning of the final approach segment. It is designated on government charts by the Maltese cross symbol for nonprecision approaches and the lightning bolt symbol for precision approaches; or when ATC directs a lower-than-published glideslope/path intercept altitude, it is the resultant actual point of the glideslope/path intercept.

final approach point (FAP). The point, applicable only to a nonprecision approach with no depicted FAF (such as an on-airport VOR), where the aircraft is established inbound on the final approach course from the procedure turn and where the final approach descent may be commenced. The FAP serves as the FAF and identifies the beginning of the final approach segment.

final controller. The controller providing information and final approach guidance during PAR and ASR approaches utilizing radar equipment.

Flight Check. A call-sign prefix used by FAA aircraft engaged in flight inspection/certification of navigational aids and flight procedures. The word "recorded" may be added as a suffix; e.g., "Flight Check 320 recorded" to indicate that an automated flight inspection is in progress in terminal areas.

flight following. *See* radar flight following.

flight level (FL). A level of constant atmospheric pressure related to a reference datum of 29.92 inches of mercury. Each is stated in three digits that represent hundreds of feet. For example, flight level (FL) 250 represents a barometric altimeter indication of 25,000 feet; FL255, an indication of 25,500 feet.

flight plan. Specified information relating to the intended flight of an aircraft that is filed orally or in writing with an FSS or an ATC facility.

flight service station (FSS). Air traffic facilities that provide pilot briefing, enroute communications, and VFR search and rescue services; assist lost aircraft and aircraft in emergency situations; relay ATC clearances; originate Notices to Airmen; broadcast aviation weather and NAS information; receive and process IFR flight plans; and monitor NAVAIDs. In addition, at selected locations, FSS's provide Enroute Flight Advisory Service (Flight Watch), take weather observations, issue airport advisories, and advise Customs and Immigration of trans-border flights.

flight test. Used after an aircraft identification, indicates an FAA flight test airplane.

Flight Watch. A shortened term for use in air/ground contacts to identify the flight service station providing Enroute Flight Advisory Service (EFAS); e.g., "Oakland Flight Watch." (*See* Enroute Flight Advisory Service.)

fly heading [*degrees*]. Informs the pilot of the heading he should fly. The pilot may have to turn to, or continue on, a specific compass direction in order to comply with the instructions. The pilot is expected to turn in the shorter direction to the heading unless otherwise instructed by ATC.

FSS. *See* flight service station.

fuel remaining. A phrase used by either pilots or controllers when relating to the fuel remaining on board until actual fuel exhaustion. When transmitting such information in response to either a controller question or pilot initiated cautionary advisory to air traffic control, pilots will state the *approximate number of minutes* the flight can continue with the fuel remaining. All reserve fuel *should be included* in the time stated, as should an allowance for established fuel gauge system error.

GCO. *See* ground controlled approach.

ground communication outlets (GCO). A means of making radio contact. Where installed (identified in the A/FD), you can access the FSS by clicking the push-to-talk button on the microphone six times; this will establish a telephone connection to the FSS briefer.

Four clicks will connect you to ATC to pick up an instrument clearance or close an IFR flight plan. Use this facility for last-minute weather briefings or to close flight plans. Do not use to file a flight plan.

ground control. At tower-controlled airports, a position in the tower responsible for controlling aircraft taxiing to and from the runways.

ground controlled approach (GCO). A radar approach system operated from the ground by ATC personnel transmitting instructions to the pilot by radio. The approach may be conducted with surveillance radar (ASR) only or with both surveillance and precision approach radar (PAR). Use of the term "GCA" by pilots is discouraged except when referring to a GCA facility. Pilots should specifically request a "PAR" approach when a precision radar approach is desired or request an "ASR" or "surveillance" approach when a nonprecision radar approach is desired.

handoff. An action taken to transfer the radar identification of an aircraft from one controller to another if the aircraft will enter the receiving controller's airspace and radio communications with the aircraft will be transferred.

have numbers. Used by pilots to inform ATC that they have received runway, wind, and altimeter information only.

Hazardous Inflight Weather Advisory Service (HIWAS). Continuous recorded hazardous inflight weather forecasts broadcasted to airborne pilots over selected VOR outlets defined as a HIWAS broadcast area.

Hazardous Weather Information. Summary of significant meteorological information (SIGMET/WS), convective significant meteorological information (convective SIGMET/WST), urgent pilot weather reports (urgent PIREP/UUA), Center weather advisories (CWA), airmen's meteorological information (AIRMET/WA) and any other weather — such as isolated thunderstorms that are rapidly developing and increasing in intensity, or low ceilings and visibilities becoming widespread and considered significant, that are not included in a current hazardous weather advisory.

HIWAS. *See* Hazardous Inflight Weather Advisory Service.

hold-short point. A point on the runway beyond which a landing aircraft with a LAHSO clearance is not authorized to proceed. This point may be located prior to an intersecting runway, taxiway, predetermined point, or approach/departure flight path.

How do you hear me? A question relating to the quality of the transmission or to determine how well the transmission is being received.

ICAO. International Civil Aviation Organization.

ident. A request for a pilot to activate the aircraft transponder identification feature. This will help the controller to confirm an aircraft identity or to identify an aircraft.

if no transmission received for *[time].* Used by ATC in radar approaches to prefix procedures that should be followed by the pilot in event of lost communications.

immediately. Used by ATC or pilots when such action compliance is required to avoid an imminent situation.

increase speed to *[speed].* *See* speed adjustment.

inflight weather advisory. *See* weather advisory.

information request (INREQ). A request originated by an FSS for information concerning an overdue VFR aircraft.

instrument departure procedure (DP). A preplanned instrument flight rule (IFR) ATC departure procedure printed for pilot use in graphic and/or textual form. DPs provide transition from the terminal to the appropriate enroute structure.

I say again. The message will be repeated.

jamming. Electronic or mechanical interference which may disrupt the display of aircraft on radar or the transmission/reception of radio communications/navigation.

known traffic. With respect to ATC clearances, means aircraft whose altitude, position, and intentions are known to ATC.

LAHSO. *See* land and hold short operations.

land and hold short operations (LAHSO). Operations that include simultaneous take-offs and landings and/or simultaneous landings, when a landing aircraft is able and is instructed by the controller to hold short of the intersecting runway/taxiway or designated hold-short point. Pilots are expected to promptly inform the controller if the hold-short clearance cannot be accepted.

land long (or long landing). Touching down on the latter part of the runway.

lifeguard. Followed by an aircraft identification, indicates an air ambulance or emergency helicopter.

local airport advisory (LAA). A service provided by flight service stations or the military at airports not serviced by an operating control tower. This service consists of providing information to arriving and departing aircraft concerning wind direction and speed, favored runway, altimeter setting, pertinent known traffic, pertinent known field conditions, airport taxi routes and traffic patterns, and authorized instrument approach procedures. This information is advisory in nature and does not constitute an ATC clearance.

local traffic. Aircraft operating in the traffic pattern or within sight of the tower, or aircraft known to be departing or arriving from flight in local practice areas, or aircraft executing practice instrument approaches at the airport.

lost communications. Loss of the ability to communicate by radio. Aircraft are sometimes referred to as NORDO (no radio). Standard pilot procedures are specified in Part 91. Radar controllers issue procedures for pilots to follow in the even of lost communications during a radar approach when weather reports indicate that an aircraft will likely encounter IFR weather conditions during the approach.

low altitude alert, check your altitude immediately. *See* safety alert.

Low Altitude Alert System (LAAS). An automated function of the TPX-42 that alerts the controller when a Mode C transponder-equipped aircraft on an IFR flight plan is below a predetermined minimum safe altitude. If requested by the pilot, LAAS monitoring is also available to VFR Mode C transponder-equipped aircraft.

maintain. Concerning altitude/flight level, the term means to remain at the altitude/flight level specified. The phrase "climb and" or "descend and" normally precedes "maintain" and the altitude assignment; e.g., "descend and maintain 5,000." Concerning other ATC instructions, the term is used in its literal sense; e.g., maintain VFR.

make short approach. Used by ATC to inform a pilot to alter his traffic pattern so as to make a short final approach.

mayday. The international radio telephony distress signal. When repeated three times, it indicates imminent and grave danger and that immediate assistance is requested.

military operations area (MOA). *See* special use airspace.

minimum fuel. Indicates that an aircraft's fuel supply has reached a state where, upon reaching the destination, it can accept little or no delay. This is not an emergency situation but merely indicates an emergency situation is possible should any undue delay occur.

missed approach. A maneuver conducted by a pilot when an instrument approach cannot be completed to a landing. The route of flight and altitude are shown on instrument approach procedure charts. A pilot executing a missed approach prior to the missed approach point (MAP) must continue along the final approach to the MAP. The pilot may climb immediately to the altitude specified in the missed approach procedure. A term used by the pilot to inform ATC that he is executing the missed approach.

MOA. *See* military operations area, special use airspace.

monitor. (When used with communication transfer) listen on a specific frequency and stand by for instructions. Under normal circumstances do not establish communications.

MULTICOM. A mobile service not open to public correspondence used to provide communications essential to conduct the activities being performed by or directed from private aircraft. "[airport name] traffic." Accessed using 122.9 MHz.

NAS. *See* National Airspace System.

National Airspace System (NAS). The common network of U.S. airspace; air navigation facilities, equipment and services, airports or landing areas; aeronautical charts, information and services; rules, regulations and procedures, technical information, and manpower and material. Included are system components shared jointly with the military.

negative. "No," or "permission not granted," or "that is not correct."

negative contact. Used by pilots to inform ATC that:

    a. Previously issued traffic is not in sight. It may be followed by the pilot's request for the controller to provide assistance in avoiding the traffic.

    b. They were unable to contact ATC on a particular frequency.

no gyro approach. A radar approach/vector provided in case of a malfunctioning gyro compass or directional gyro. Instead of providing the pilot with headings to be flown, the controller observes the radar track and issues control instructions "turn right/left" or "stop turn" as appropriate.

no gyro vector. *See* no gyro approach.

NORDO. No radio. *See* lost communications.

numerous targets vicinity *[location]*. A traffic advisory issued by ATC to advise pilots that targets on the radar scope are too numerous to issue individually. *See* traffic advisories.

off course. A term used to describe a situation where an aircraft has reported a position fix or is observed on radar at a point not on the ATC-approved route of flight.

off-route vector. A vector by ATC which takes an aircraft off a previously assigned route. Altitudes assigned by ATC during such vectors provide required obstacle clearance.

on-course. Used to indicate that an aircraft is established on the route centerline. Used by ATC to advise a pilot making a radar approach that his aircraft is lined up on the final approach course.

option approach. An approach requested and conducted by a pilot which will result in either a touch-and-go, missed approach, low approach, stop-and-go, or full stop landing.

out. The conversation is ended and no response is expected.

outer area (associated with Class C airspace). Nonregulatory airspace surrounding designated Class C airspace airports wherein ATC provides radar vectoring and sequencing on a full-time basis for all IFR and participating VFR aircraft. The service provided in the outer area is called Class C service which includes: IFR/IFR—standard IFR separation; IFR/VFR—traffic advisories and conflict resolution; VFR/VFR—traffic advisories; and, as appropriate, safety alerts. The normal radius will be 20 nautical miles with some variations based on site-specific requirements. The outer area extends outward from the primary Class C airspace airport and extends from the lower limits of radar/radio coverage up to the ceiling of the approach control's delegated airspace excluding the Class C charted area and other airspace as appropriate.

over. My transmission is ended; I expect a response.

pan-pan-pan. The international radio-telephony urgency signal. When repeated three times, indicates uncertainty or alert followed by the nature of the urgency.

PAR. *See* precision approach radar.

pilot briefing. A service provided by the FSS to assist pilots in flight planning. Briefing items may include weather information, NOTAMs, military activities, flow control information, and other items as requested.

pilot-in-command. The pilot responsible for the operation and safety of an aircraft during flight time.

pilot's discretion. When used in conjunction with altitude assignments, means that ATC has offered the pilot the option of starting climb or descent whenever he wishes and conducting the climb or descent at any rate he wishes. He may temporarily level off at any intermediate altitude. However, once he has vacated an altitude, he may not return to that altitude.

pilot weather report (PIREP). A report of meteorological phenomena encountered by aircraft in flight.

PIREP. *See* pilot weather report.

position report. A report over a known location as transmitted by an aircraft to ATC.

practice instrument approach. An instrument approach procedure conducted by a VFR or IFR aircraft for the purpose of pilot training or proficiency demonstrations.

precision approach radar (PAR). Radar equipment in some ATC facilities operated by the FAA and/or the military services at joint-use civil/military locations and separate military installations to detect and display azimuth, elevations, and range of aircraft on the final approach course to a runway. This equipment may be used to monitor certain nonradar approaches, but is primarily used to conduct a precision instrument approach (PAR) wherein the controller issues guidance instructions to the pilot based on the aircraft's position in relation to the final approach course (azimuth), the glide-path (elevation), and the distance (range) from the touchdown point on the runway as displayed on the radar scope. *Note:* The abbreviation PAR is also used to denote preferential arrival routes in ARTCC computers.

preferential routes. Preferential routes (PDRs, PARs, and PDARs) are adapted in ARTCC computers to accomplish inter/intrafacility controller coordination and to ensure that flight data is posted at the proper control positions. Locations having a need for these specific inbound and outbound routes normally publish such routes in local facility bulletins, and their use by pilots minimizes flight plan route amendments. When the workload or traffic situation permits, controllers normally provide radar vectors or assign requested routes to minimize circuitous routing. Preferential routes are usually confined to one ARTCC's area and are referred to by the following names or acronyms:

   a. Preferential departure route (PDR). A specific departure route from an airport or terminal area to an enroute point where there is no further need for flow control. It may be included in a instrument departure procedure (DP) or preferred IFR route.

   b. Preferential arrival route (PAR). A specific arrival route from an appropriate enroute point to an airport or terminal area. It may be included in a standard terminal arrival (STAR) or preferred IFR route. The abbreviation PAR is used primarily within the ARTCC and should not be confused with the abbreviation for precision approach radar.

   c. Preferential departure and arrival route (PDAR). A route between two terminals that are within or immediately adjacent to one ARTCC's area. PDARs are not synonymous with preferred IFR routes but may be listed as such, as they do accomplish essentially the same purpose. (*See* preferred IFR routes)

preferred IFR routes. Routes established between busier airports to increase system efficiency and capacity. They normally extend through one or more ARTCC areas and are designed to achieve balanced traffic flows among high-density terminals. IFR clearances are issued on the basis of these routes except when severe weather avoidance procedures or other factors dictate otherwise. Preferred IFR routes are listed in the Airport/Facility Directory. If a flight is planned to or from an area having such routes but the departure or arrival point is not listed in the Airport/Facility Directory, pilots may use that part of a preferred IFR Route which is appropriate for the departure or arrival point that is listed. Preferred IFR Routes are correlated with DPs and STARs and may be defined by airways, jet routes, direct routes between NAVAIDs, waypoints, NAVAID radials/ DME, or any combinations thereof. (*See* instrument departure procedure) (*See* standard terminal arrival) (*See* preferential routes) (Refer to Airport/Facility Directory) (Refer to *Notices to Airmen Publication*)

progressive taxi. Precise taxi instructions given to a pilot unfamiliar with the airport or issued in stages as the aircraft proceeds along the taxi route.

prohibited area. *See* special use airspace.

radar advisory. The provision of advice and information based on radar observations.

radar approach. An instrument approach procedure that uses precision approach radar (PAR) or airport surveillance radar (ASR).

radar approach control facility. A terminal ATC facility that uses radar and nonradar capabilities to provide approach control services to aircraft arriving, departing, or transiting airspace controlled by the facility. Provides radar ATC services to aircraft operating in the vicinity of one or more civil and/or military airports in a terminal area. The facility may provide services of a ground controlled approach (GCA); i.e., ASR and PAR approaches. A radar approach control facility may be operated by FAA, USAF, US Army, USN, USMC, or jointly by FAA and a military service. Specific facility nomenclatures are used for administrative purposes only and are related to the physical location of the facility and the operating service generally as follows:

    a. Army radar approach control (ARAC) (Army).

    b. Radar air traffic control facility (RATCF) (Navy/FAA).

    c. Radar approach control (RAPCON) (Air Force/FAA).

    d. Terminal radar approach control (TRACON) (FAA).

    e. Air traffic control tower (ATCT) (FAA). (Only those towers that are delegated approach control authority.)

radar contact. Used by ATC to inform an aircraft that it is identified on the radar display and radar flight following will be provided until radar identification is terminated. Radar service may also be provided within the limits of necessity and capability. When a pilot is informed of "radar contact," he automatically discontinues reporting over compulsory reporting points. The term used to inform the controller that the aircraft is identified and approval is granted for the aircraft to enter the receiving controller's airspace.

radar contact lost. Used by ATC to inform a pilot that radar data used to determine the aircraft's position is no longer being received, or is no longer reliable and radar service is no longer being provided. The loss may be attributed to several factors including the aircraft merging with weather or ground clutter, the aircraft operating below radar line of sight coverage, the aircraft entering an area of poor radar return, failure of the aircraft transponder, or failure of the ground radar equipment.

radar flight following. The observation of the progress of radar identified aircraft, whose primary navigation is being provided by the pilot, wherein the controller retains and correlates the aircraft identity with the appropriate target or target symbol, displayed on the radar scope.

radar service terminated. Used by ATC to inform a pilot that he will no longer be provided any of the services that could be received while in radar contact. Radar service is automatically terminated, and the pilot is not advised in the following cases:

   a. An aircraft cancels its IFR flight plan, except within Class B airspace, Class C airspace, a TRSA, or where basic radar service is provided.

   b. An aircraft conducting an instrument, visual, or contact approach has landed or has been instructed to change to advisory frequency.

   c. An arriving VFR aircraft, receiving radar service to a tower-controlled airport within Class B airspace, Class C airspace, a TRSA, or where sequencing service is provided, has landed; or to all other airports, is instructed to change to tower or advisory frequency.

   d. An aircraft completes a radar approach.

radio.
   a. A device used for communication.

   b. Used to refer to a flight service station (FSS); e.g., "Seattle Radio" is used to call Seattle FSS.

RCO. *See* remote communications outlet.

read back. Repeat my message back to me.

reading the mail. Listening on a frequency.

receiving controller. A controller/facility receiving control of an aircraft from another controller/facility.

reduce speed to [speed]. *See* speed adjustment.

remote communications outlet (RCO). An unmanned communications facility remotely controlled by air traffic personnel. RCOs serve FSS's. Remote transmitter/receivers (RTRs) serve terminal ATC facilities. An RCO or RTR may be UHF or VHF and will extend the communication range of the air traffic facility. There are several classes of RCOs and RTRs. The class is determined by the number of transmitters or receivers. Classes A through G are used primarily for air/ground purposes. RCO and RTR class O facilities are nonprotected outlets subject to undetected and prolonged outages. RCO (O)'s and

RTR (O)'s were established for the express purpose of providing ground-to-ground communications between air traffic control specialists and pilots located at a satellite airport for delivering enroute clearances, issuing departure authorizations, and acknowledging instrument flight rules cancellations or departure/landing times. As a secondary function, they may be used for advisory purposes whenever the aircraft is below the coverage of the primary air/ground frequency.

report. Used to instruct pilots to advise ATC of specified information; e.g., "Report passing Hamilton VOR."

request full route clearance. Used by pilots to request that the entire route of flight be read verbatim in an ATC clearance. Such request should be made to preclude receiving an ATC clearance based on the original filed flight plan when a filed IFR flight plan has been revised by the pilot, company, or operations prior to departure.

restricted area. *See* special use airspace.

resume own navigation. Used by ATC to advise a pilot to resume his own navigational responsibility. It is issued after completion of a radar vector, or when radar contact is lost while the aircraft is being radar-vectored.

resume normal speed. Used by ATC to advise a pilot that previously-issued speed control restrictions are deleted. An instruction to "resume normal speed" does not delete speed restrictions that are applicable to published procedures of upcoming segments of flight, unless specifically stated by ATC. This does not relieve the pilot of those speed restrictions that are applicable to 14 CFR §91.117.

roger. I have received all of your last transmission. It should not be used to answer a question requiring a yes or a no answer.

runway heading. The magnetic direction that corresponds with the runway centerline extended, not the painted runway number. When cleared to "fly or maintain runway heading," pilots are expected to fly or maintain the heading that corresponds with the extended centerline of the departure runway. Drift correction shall not be applied; e.g., Runway 4, actual magnetic heading of the runway centerline 044, fly 044.

runway in use/active runway/duty runway. Any runway or runways currently being used for takeoff or landing. When multiple runways are used, they are all considered active runways.

safety alert. A safety alert issued by ATC to aircraft under their control if ATC is aware the aircraft is at an altitude which, in the controller's judgment, places the aircraft in unsafe proximity to terrain, obstructions, or other aircraft. The controller may discontinue the issuance of further alerts if the pilot advises he is taking action to correct the situation or has the other aircraft in sight.

    a. Terrain/obstruction alert — A safety alert issued by ATC to aircraft under their control if ATC is aware the aircraft is at an altitude which, in the controller's judgment, places the aircraft in unsafe proximity to terrain/obstructions, e.g., "Low altitude alert, check your altitude immediately."

b. Aircraft conflict alert — A safety alert is issued by ATC to aircraft under their control if ATC is aware of an aircraft that is not under their control at an altitude which, in the controller's judgment, places both aircraft in unsafe proximity to each other. With the alert, ATC will offer the pilot an alternate course of action when feasible; e.g., "Traffic alert, advise you turn right heading zero niner zero or climb to eight thousand immediately."

The issuance of a safety alert is contingent upon the capability of the controller to have an awareness of an unsafe condition. The course of action provided will be predicated on other traffic under ATC control. Once the alert is issued, it is solely the pilot's prerogative to determine what course of action, if any, he will take.

same direction aircraft. Aircraft are operating in the same direction when:

    a. They are following the same track in the same direction; or

    b. Their tracks are parallel and the aircraft are flying in the same direction; or

    c. Their tracks intersect at an angle of less than 45 degrees.

say again. Used to request a repeat of the last transmission. Usually specifies transmission or portion thereof not understood or received; e.g., "Say again all after ABRAM VOR."

say altitude. Used by ATC to ascertain an aircraft's specific altitude/flight level. When the aircraft is climbing or descending, the pilot should state the indicated altitude rounded to the nearest 100 feet.

say heading. Used by ATC to request an aircraft heading. The pilot should state the actual heading of the aircraft.

self announce. At nontowered airports, pilots identify position when about 10 miles from the destination airport, giving identification, position in relation to the airport, altitude, and intentions.

SIGMET. A weather advisory issued concerning weather significant to the safety of all aircraft. SIGMET advisories cover severe and extreme turbulence, severe icing, and widespread dust or sandstorms that reduce visibility to less than 3 miles.

significant meteorological information. *See* SIGMET.

speak slower. Used in verbal communications as a request to reduce speech rate.

special use airspace. Airspace of defined dimensions identified by an area on the surface of the earth wherein activities must be confined because of their nature and/or wherein limitations may be imposed upon aircraft operations that are not a part of those activities. Types of special use airspace are:

    a. Alert area — Airspace that may contain a high volume of pilot-training activities or an unusual type of aerial activity, neither of which is hazardous to aircraft. Alert areas are depicted on aeronautical charts for the information of nonparticipating pilots. All activities within an alert area are conducted in accordance with Federal Aviation Regulations, and pilots of participating aircraft as well as pilots transiting the area are equally responsible for collision avoidance.

Continued

b. Controlled firing area — Airspace wherein activities are conducted under conditions so controlled as to eliminate hazards to nonparticipating aircraft and to ensure the safety of persons and property on the ground.

c. Military operations area (MOA) — Airspace established outside of Class A airspace area to separate or segregate certain non-hazardous military activities from IFR traffic and to identify for VFR traffic where these activities are conducted.

d. Prohibited area — Airspace designated under Part 73 within which no person may operate an aircraft without the permission of the using agency.

e. Restricted area — Airspace designated under Part 73, within which the flight of aircraft, while not wholly prohibited, is subject to restriction. Most restricted areas are designated joint use and IFR/VFR operations in the area may be authorized by the controlling ATC facility when it is not being utilized by the using agency. Restricted areas are depicted on enroute charts. Where joint use is authorized, the name of the ATC controlling facility is also shown.

f. Warning area — A warning area is airspace of defined dimensions extending from 3 nautical miles outward from the coast of the United States, that contains activity that may be hazardous to nonparticipating aircraft. The purpose of such warning area is to warn nonparticipating pilots of the potential danger. A warning area may be located over domestic or international waters or both.

special VFR operations (SVFR). Aircraft operating in accordance with clearances within Class B, C, D, and E surface areas in weather conditions less than the basic VFR weather minima. Such operations must be requested by the pilot and approved by ATC.

speed adjustment. An ATC procedure used to request pilots to adjust aircraft speed to a specific value for the purpose of providing desired spacing. Pilots are expected to maintain a speed of plus or minus 10 knots or 0.02 Mach number of the specified speed. Examples of speed adjustments are:

a. "Increase/reduce speed to Mach point *[number]*."

b. "Increase/reduce speed to *[speed in knots]*" or "Increase/reduce speed *[number of knots]* knots."

squawk *[mode, code, function]*. Activate specific modes/codes/functions on the aircraft transponder, e.g., "Squawk three/Alfa, two one zero five, low."

squelch. A circuit in a radio receiver that keeps the volume down when no signal is being received. As soon as a signal is received, the squelch circuit allows it to come through loud enough to be comfortably heard.

standard terminal arrival (STAR). A preplanned instrument flight rule (IFR) air traffic control arrival procedure published for pilot use in graphic and/or textual form. STARs provide transition from the enroute structure to an outer fix or an instrument approach fix/arrival waypoint in the terminal area.

stand by. Means the controller or pilot must pause for a few seconds, usually to attend to other duties of a higher priority. Also means to wait as in "stand by for clearance."

The caller should reestablish contact if a delay is lengthy. "Stand by" is not an approval or denial.

STAR. *See* standard terminal arrival.

stop altitude squawk. Used by ATC to inform an aircraft to turn-off the automatic altitude reporting feature of its transponder. It is issued when the verbally reported altitude varies 300 feet or more from the automatic altitude report.

stop and go. A procedure wherein an aircraft will land, make a complete stop on the runway, and then commence a takeoff from that point.

stop squawk *[mode or code].* Used by ATC to tell the pilot to turn specified functions of the aircraft transponder off.

Tango. Used after an aircraft identification, indicates an air taxi operator.

taxi into position and hold. Used by ATC to inform a pilot to taxi onto the departure runway in takeoff position and hold. It is not authorization for takeoff. It is used when takeoff clearance cannot immediately be issued because of traffic or other reasons.

TCAS. *See* traffic alert and collision avoidance system.

terminal radar service area (TRSA). Airspace surrounding designated airports wherein ATC provides radar vectoring, sequencing, and separation on a full-time basis for all IFR and participating VFR aircraft. TRSAs are depicted on VFR aeronautical charts. Pilot participation is urged but is not mandatory.

terminal VFR radar service. A national program instituted to extend the terminal radar services provided instrument flight rules (IFR) aircraft to visual flight rules (VFR) aircraft. The program is divided into four types of service referred to as basic radar service, terminal radar service area (TRSA) service, Class B service and Class C service. The type of service provided at a particular location is contained in the Airport/Facility Directory.

    a. Basic radar service—These services are provided for VFR aircraft by all commissioned terminal radar facilities. Basic radar service includes safety alerts, traffic advisories, limited radar vectoring when requested by the pilot, and sequencing at locations where procedures have been established for this purpose and/or when covered by a letter of agreement. The purpose of this service is to adjust the flow of arriving IFR and VFR aircraft into the traffic pattern in a safe and orderly manner, and to provide traffic advisories to departing VFR aircraft.

    b. TRSA service—This service provides, in addition to basic radar service, sequencing of all IFR and participating VFR aircraft to the primary airport and separation between all participating VFR aircraft. The purpose of this service is to provide separation between all participating VFR aircraft and all IFR aircraft operating within the area defined as a TRSA.

    c. Class C service—This service provides, in addition to basic radar service, approved separation between IFR and VFR aircraft, sequencing of VFR aircraft, and sequencing of VFR arrivals to the primary airport.

Continued

d. Class B service — This service provides, in addition to basic radar service, approved separation of aircraft based on IFR, VFR, and/or weight, and sequencing of VFR arrivals to the primary airport(s).

that is correct. The understanding you have is right.

touch-and-go. An operation by an aircraft that lands and departs on a runway without stopping or exiting the runway.

tower. A terminal facility that uses air/ground communications, visual signaling, and other devices to provide ATC services to aircraft operating in the vicinity of an airport or on the movement area. Authorizes aircraft to land or takeoff at the airport controlled by the tower or to transit the Class D airspace area regardless of flight plan or weather conditions (IFR or VFR). A tower may also provide approach control services (radar or nonradar).

tower enroute control service. The control of IFR enroute traffic within delegated airspace between two or more adjacent approach control facilities. This service is designed to expedite traffic and reduce control and pilot communication requirements.

traffic. A term used by ATC to refer to one or more aircraft. A term used by a controller to transfer radar identification of an aircraft to another controller for the purpose of coordinating separation action. Traffic is normally issued: 1) in response to a handoff, 2) in anticipation of a handoff, or 3) in conjunction with a request for control of an aircraft.

traffic advisories. Advisories issued to alert pilots to other known or observed air traffic that may be in such proximity to the position or intended route of flight of their aircraft to warrant their attention. Such advisories may be based on:

a. Visual observation.

b. Observation of radar identified and non-identified aircraft targets on an ATC radar display, or

c. Verbal reports from pilots or other facilities.

Notes: The word "traffic" followed by additional information, if known, is used to provide such advisories; e.g., "Traffic, 2 o'clock, one zero miles, southbound, eight thousand." Traffic advisory service will be provided to the extent possible depending on higher priority duties of the controller or other limitations; e.g., radar limitations, volume of traffic, frequency congestion, or controller workload. Radar/nonradar traffic advisories do not relieve the pilot of his responsibility to see and avoid other aircraft. Pilots are cautioned that there are many times when the controller is not able to give traffic advisories concerning all traffic in the aircraft's proximity; in other words, when a pilot requests or is receiving traffic advisories, he should not assume that all traffic will be issued.

traffic alert *[aircraft call sign]*, turn *[left/right]* immediately, *[climb/descend]* and maintain *[altitude]*. *See* safety alert.

traffic alert and collision avoidance system. An airborne collision avoidance system based on radar beacon signals that operates independently of ground-based equipment. TCAS-I generates traffic advisories only. TCAS-II generates traffic advisories, and resolution (collision avoidance) advisories in the vertical plane.

traffic in sight. Used by pilots to inform a controller that previously issued traffic is in sight.

traffic no factor. Indicates that the traffic described in a previously issued traffic advisory is no factor.

traffic no longer observed. Indicates that the traffic described in a previously issued traffic advisory is no longer depicted on radar, but may still be a factor.

transceiver. A piece of radio communications equipment in which all of the circuits for both the receiver and transmitter are contained in the same housing.

transfer of control. That action whereby the responsibility for the separation of an aircraft is transferred from one controller to another.

transmitting in the blind. A transmission from one station to other stations in circumstances where two-way communication cannot be established, but where it is believed that the called stations may be able to receive the transmission.

TRSA. *See* terminal radar service area.

two-way radio communications failure. *See* lost communications.

unable. Indicates inability to comply with a specific instruction, request, or clearance.

unfamiliar. Indicates you are in an area where you are unaccustomed to the surroundings, local landmarks, airport environment, etc.

UNICOM. A non-government communication facility that may provide airport information at certain airports. Locations and frequencies of UNICOMs are shown on aeronautical charts and publications. "*[airport name]* UNICOM."

urgency. A condition of being concerned about safety and of requiring timely but not immediate assistance; a potential distress condition.

vector. A heading issued to an aircraft to provide navigational guidance by radar.

verify. Request confirmation of information; e.g., "verify assigned altitude."

verify specific direction of takeoff (or turns after takeoff). Used by ATC to ascertain an aircraft's direction of takeoff and/or direction of turn after takeoff. It is normally used for IFR departures from an airport not having a control tower. When direct communication with the pilot is not possible, the request and information may be relayed through an FSS, dispatcher, or by other means.

VFR corridor. Airspace through Class B airspace, with defined vertical and lateral boundaries, in which aircraft may operate without an ATC clearance or communication with air traffic control.

VFR flyway. A general flight path not defined as a specific course, for use by pilots in planning flights into, out of, through or near complex terminal airspace to avoid Class B airspace. An ATC clearance is *not* required to fly these routes.

VFR not recommended. An advisory provided by a flight service station to a pilot during a preflight or inflight weather briefing that flight under visual flight rules is not recommended. To be given when the current and/or forecast weather conditions are at or below VFR minimums. It does not abrogate the pilot's authority to make his own decision.

VFR-on-Top. ATC authorization for an IFR aircraft to operate in VFR conditions at any appropriate VFR altitude (as specified in regulations and as restricted by ATC). A pilot receiving this authorization must comply with the VFR visibility, distance from cloud criteria, and the minimum IFR altitudes specified in Part 91. The use of this term does not relieve controllers of their responsibility to separate aircraft in Class B and Class C airspace or TRSAs.

VFR transition routes (Class B airspace). A specific flight course depicted on a Terminal Area Chart (TAC) for transiting a specific Class B airspace. These routes include specific ATC-assigned altitudes, and pilots must obtain an ATC clearance prior to entering Class B airspace on the route.

visual approach. An approach conducted on an instrument flight rules (IFR) flight plan that authorizes the pilot to proceed visually and clear of clouds to the airport. The pilot must, at all times, have either the airport or the preceding aircraft in sight. This approach must be authorized and under the control of the appropriate air traffic control facility. Reported weather at the airport must be ceiling at or above 1,000 feet and visibility of 3 miles or greater.

warning area. *See* special use airspace.

weather advisory. In aviation weather forecast practice, an expression of hazardous weather conditions not predicted in the area forecast, as they affect the operation of air traffic and as prepared by the NWS.

when able. When used in conjunction with ATC instructions, gives the pilot the latitude to delay compliance until a condition or event has been reconciled. Unlike "pilot discretion," when instructions are prefaced "when able," the pilot is expected to seek the first opportunity to comply. Once a maneuver has been initiated, the pilot is expected to continue until the specifications of the instructions have been met. "When able" should not be used when expeditious compliance is required.

wilco. I have received your message, understand it, and will comply with it.

words twice. As a request: "Communication is difficult. Please say every phrase twice." As information: "Since communications are difficult, every phrase in this message will be spoken twice."